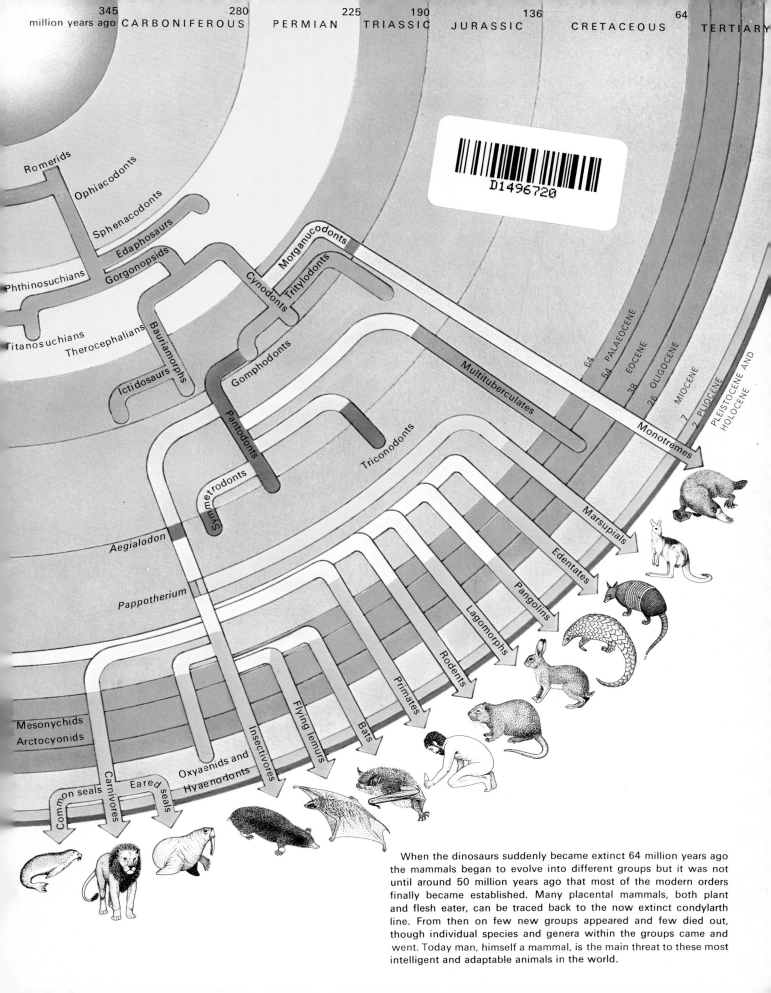

million years ago CARBONIFEROUS   PERMIAN   TRIASSIC   JURASSIC   CRETACEOUS   TERTIARY
345   280   225   190   136   64

Romerids
Ophiacodonts
Sphenacodonts
Edaphosaurs
Gorgonopsids
Phthinosuchians
Morganucodonts
Cynodonts Tritylodonts
Titanosuchians
Therocephalians
Bauriamorphs
Gomphodonts
Ictidosaurs
Multituberculates
Pantodonts
Symmetrodonts
Triconodonts
Aegialodon
Pappotherium

64 PALAEOCENE
54 EOCENE
38 OLIGOCENE
26 MIOCENE
7   PLIOCENE
2   PLEISTOCENE AND HOLOCENE

Monotremes
Marsupials
Edentates
Pangolins
Lagomorphs
Rodents
Primates
Bats
Flying lemurs
Insectivores
Oxyaenids and Hyaenodonts
Mesonychids
Arctocyonids
Common seals
Carnivores
Eared seals

When the dinosaurs suddenly became extinct 64 million years ago the mammals began to evolve into different groups but it was not until around 50 million years ago that most of the modern orders finally became established. Many placental mammals, both plant and flesh eater, can be traced back to the now extinct condylarth line. From then on few new groups appeared and few died out, though individual species and genera within the groups came and went. Today man, himself a mammal, is the main threat to these most intelligent and adaptable animals in the world.

THE EVOLUTION OF
The Mammals

THE EVOLUTION OF
The Mammals

by L.B. Halstead

With eight double-page paintings by Sergio

Peter Lowe

NOTE ON MAMMAL NAMES

Living things are divided into two large, general groups, the plant kingdom and the animal kingdom. These are sub-divided into smaller and smaller categories – phyla, classes, orders and families. The families each contain several genera (genus means literally race) and each genus contains a group of species which are closely related and often look alike.

These divisions reflect the way animals have evolved from similar ancestors into distinct species. The differences between species have evolved most recently; differences between orders are older and between classes and phyla more ancient still.

An animal's latin name is made up of its genus and species and distinguishes it from every other type of animal. If they belong to the same species, animals can breed with one another and produce fertile young. If they are from different species but the same genus they can sometimes breed successfully, but their young will be infertile.

Fossil animals, like living ones, are all given names but are included in the same genus and species as modern types only if they have the same main characteristics. Other fossils may be given a different genus name but be included in the wider relationship of the family. A few extinct genera and species belonged to orders which have died out altogether and cannot be directly linked with any modern groups.

Some of the names of early mammals will be unfamiliar at first but in fact most have quite simple meanings. Several end in '-*therium*'. This means literally 'beast'. Others end in '-*dont*' – meaning teeth. A hyaenodont is simply an animal with teeth like a hyaena's. Names beginning with '*Mega*-' mean that the animal is particularly large – *Megatherium* means 'large beast'.

Previous page

Fossil evidence proves that 35 million years ago a fight like this actually took place in North America between an early sabre-tooth cat *Eusmilus* and a biting cat *Nimravus*. We know that the deep wound made in *Nimravus'* head by *Eusmilus'* dagger-like teeth later healed. But we cannot tell whether *Eusmilus* also survived.

For Jenny

British Library Cataloguing in Publication Data
Halstead, Lambert Beverly
  The evolution of the mammals.
  1. Mammals – Evolution – Juvenile literature
  I. Title
  599'.03'8   QL708.3
  ISBN 0 85654 030 7

Printed in Italy by New Interlitho Spa

Contents

Introduction

By far the most familiar and widespread animals and to date the most successful throughout the history of the world are human beings. From the Arctic wastes to equatorial rain forests and the blistering heat of the deserts there are people. The ability to live in such a wide range of conditions is not confined to human beings alone but is characteristic of the entire group to which man belongs – the mammals. This is because all these animals have acquired a constant high internal temperature which makes them to a great extent independent of the surrounding conditions.

Perhaps the most fundamental feature of the mammals compared with all other known living things is their greater intelligence and their ability to learn over long periods. When the young are born they are cared for by their parents and are fed on their mother's milk. The very name *mammal* comes from the Latin word for breasts *mammae*, which in its turn is the universal term for mother. At its mother's breast the young mammal begins the long period of education which prepares it for its later independence in life.

For 64 million years the world has been dominated by mammals and since the beginning of the Ice Age two and a half million years ago a single species of mammal, man, has ruled the living world. The evolution of the mammals, however, did not begin with the extinction of the dinosaurs 64 million years ago. The great dynasty of paramammals which gave rise to the mammals originated in the tropical coal swamps some 300 million years ago and by 200 million years ago the first true mammals had appeared. Then came their dramatic eclipse as the dinosaurs took over. For the next 140 million years tiny, furry mammals remained inconspicuous, venturing out to feed only at night. At the end of the Age of Dinosaurs the world climate changed and plant food became less plentiful. The giant dinosaurs with their need for vast quantities of food were thrust into a crisis from which they were unable to recover. It was in just such difficult times that the mammals came into their own.

The Age of Mammals began, yet a visitor to Earth 64 million years ago would have found it hard to believe that from creatures such as the shy, shrew-like insect eaters and squirrel-like plant eaters of the night would evolve beings capable of harnessing nuclear energy and travelling beyond the confines of the planet Earth. This book is the story of those most remarkable of all living things – the mammals.

Beverly Halstead

What is a Mammal?

For the last 64 million years, mammals have been the dominant animals on earth. They have spread all over the world, adapting to habitats as different as oceans and deserts. They include herbivores, carnivores, insectivores and omnivores; night hunters, scavengers and peaceful daytime grazers and browsers. There are mammals that fly and others that spend their whole life in the water. In size they range from tiny shrews to giant whales and elephants. And they include our own species, *Homo sapiens*.

The definition of a mammal is an animal that suckles its young, but they have other characteristics in common which set them apart from other animal groups.

Temperature control

The first mammals appeared over 200 million years ago, when dinosaurs were the dominant animals. For over 140 million years they seem to have made little impact on life, remaining small, shrew-like insect-eaters. They managed to survive because they occupied a place in the environment that insect-eating reptiles could not use: they were active at night.

During the daytime lizards and other similar reptiles can control the internal temperature of their bodies (though only within very narrow limits) by the pattern of their behaviour. When they are too hot they seek the shade, when too cool, the warmth of the sunlight. During the night they become sluggish and go into a kind of torpor. This type of animal is called an ectotherm because its body temperature is governed by outside conditions. In marked contrast, mammals maintain a constant high internal temperature by burning up their food rapidly. They are said to have a high metabolic rate and are called endotherms because their temperature is controlled from within themselves.

If a high metabolic rate is to be properly effective, the animal needs some way of insulating its body so that the heat does not escape. Almost all mammals have a furry or hairy covering of some kind which traps a layer of air around their body and prevents it from either cooling down or heating up too much. Birds, the other successful group of warm-blooded animals, use feathers in the same way.

Although fur and hair are effective insulators, mammals still need to cool down from time to time. They do this by sweating and panting. As water is excreted in the form of sweat it evaporates and the body is cooled. Other types of animals have ways of cooling themselves, but the development of sweat glands in the skin was important in mammals as it led eventually to the evolution of their special suckling glands.

The fundamental feature common to all mammals is that the young are fed on milk produced by the mother. The young animals are also cared for by their parents and while they are dependent they gradually learn all the skills they need to survive as adults. Horses are not fully mature until they are six years old. Man, the dominant animal of today, has the longest learning period of all.

A mammal's breast contains milk-producing glands, separated by fibrous and fatty tissue. Ducts leading from the glands open out to form reservoirs and the milk finally oozes out through minute pores on the nipple.

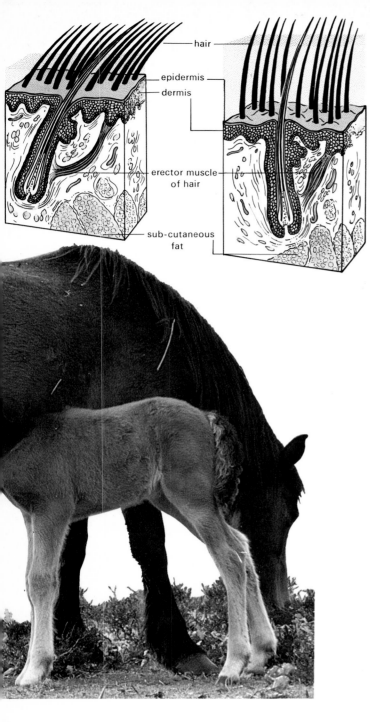

hair

epidermis

dermis

erector muscle
of hair

sub-cutaneous
fat

A mammal's fur or hair helps to maintain its constant high internal temperature. When it is warm the hairs lie flat. When it is cold, a special muscle pulls the hairs erect, trapping a larger layer of air. This acts as an insulator, preventing heat loss from the surface of the skin. Sweat glands provide a mechanism for cooling down when the animal gets over-heated.

The neocortex is the part of the brain concerned with intelligence and it is much more developed in mammals than in any other animal. Because it is folded in on itself, a large surface area can be contained inside quite a small volume.

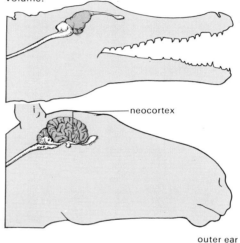

neocortex

Mammals have three bones in the middle ear which conduct sound vibrations from the ear-drum to the fluid-filled cochlea. The vibrations increase as they are passed from bone to bone and this sound amplification system gives mammals an extremely acute sense of hearing.

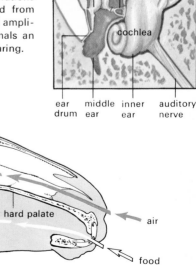

semi-circular
canals

outer ear   stapes

bone

incus

malleus

cochlea

ear   middle inner   auditory
drum   ear   ear   nerve

molar

premolar   canine

incisor

Mammals have different types of teeth for performing different tasks. The incisors are for food gathering, the cheek teeth for grinding and slicing food in the mouth. The canines are weapons and are most highly developed in flesh-eating mammals.

soft palate   hard palate

air

food

Only mammals have a secondary palate separating air and food passages so that they can chew and breathe at the same time. The hard part stays rigid even when food is pressed against it; the soft flexible part closes the air intake as food is actually swallowed.

Because they were already warm blooded, the early mammals could be active at night, when the cold-blooded reptiles were asleep. The mammals were insignificant compared to the great variety of dinosaurs but they dominated the night life of the time.

Breathing and eating

To keep its internal temperature at the right level, an animal needs a constant supply of oxygen and food. Mammals have a secondary palate which separates the air passage from the food passage so that they can hold food in their mouth and partially prepare it for digestion without interrupting breathing. The secondary palate makes it possible for young animals to suckle at the breast of their mother and continue to breathe as they do so. Mammals also have very efficient lungs, divided from their digestive organs by a diaphragm.

Teeth may not seem such an obvious characteristic as fur, but in fact they are often used to distinguish mammals from other animals. Animals without secondary palates use their teeth mainly to prevent food from escaping from their mouth. When an animal has a secondary palate, it can use its teeth to break down or chew the food. Mammals' teeth became gradually pointed (for crushing and puncturing) or ridged (for shearing). Moreover, the teeth in different parts of the jaw became adapted for performing different tasks. The front teeth, the incisors, pluck and pull off pieces of food; the long, sharp canine teeth are used for tearing and stabbing; the cusped cheek teeth are for chewing. By comparison, reptile teeth are basically the same wherever they grow in the jaw. Mammal jaws are efficiently adapted for obtaining food and the single lower jaw bone makes them stronger than reptile jaws.

Reproduction and care of the young

Most animals, including insects, fish, birds and reptiles, lay eggs. In all except one group of mammals, the fertilized egg remains within the mother's body. The yolk (which in birds and reptiles provides food for the developing embryo) is small and there is a special structure, the placenta, which links tissues from the mother and the embryo so that their blood supplies come into very close contact. The embryo receives food and gets rid of its waste products via the placenta. In this way the mother can provide her developing young with all the food it needs and it is able to grow in a protected environment during the most vulnerable period of its life.

After the young mammal has been born, the mother continues to feed it with her milk. Suckling is one of the

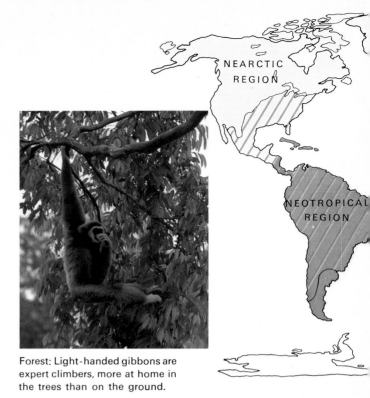

Forest: Light-handed gibbons are expert climbers, more at home in the trees than on the ground.

Desert: Camels have many special features which enable them to survive in hot, dry climates.

basic features of mammals and their survival throughout the Age of Dinosaurs and later success, was to a large extent due to this unique method of feeding. Young mammals are not only fed but also cared for by their mother. Other animals, too, show parental care: crocodiles, for example, guard their eggs and the newly-hatched young are protected. Birds incubate their eggs and feed and care for the young until they can leave the nest and fend for themselves. Young mammals, however, remain dependent on their mother for a considerable period and it is some time before they are capable of feeding themselves. While they are being cared for by their parents, young mammals go through an extended period of learning. As a result of interaction with both parents and siblings (brothers and sisters) they gradually acquire the skills necessary for survival. This long period of education is characteristic of mammals and the most successful mammals (such as man) have the most extended learning periods.

Ocean: Dolphins are perfectly adapted to their ocean environment.

Air: Fruit bat with young. Bats are the only mammals that can fly.

Savannah: Cheetahs prey on the fast running deer and antelopes of the African grasslands.

Arctic: A thick layer of fat and a furry coat insulate this harp seal pup from the cold.

Marsupials

Marsupials and Monotremes

Land mammals live in every continent except Antarctica. Monotremes are found only in Australasia, marsupials in Australasia and the Americas. Eurasia, Africa, North and South America are dominated by placental mammals, which have adapted to all kinds of environments from tropical forest to dry deserts. Marine mammals, the whales and seals, live in all the oceans including the cold Antarctic.

Mountain: Sure-footed ibexes live high up near the snow-line for most of the year.

The brain and the senses

Mammals not only have a long learning period: they also have a well developed memory and are capable of learning from experience. Even more remarkably, some have the ability to deal with problems they have not experienced before.

The part of the brain which is concerned with these activities is only properly developed in mammals: it is called the neopallium or neocortex. In the more advanced mammals, this area is made much bigger by infolding – to such an extent that it overgrows all the other parts of the brain. The mammals with the most highly developed neocortex areas of all are dolphins and people.

The evolution of the brain, so important to the evolution of mammals as a successful group of animals, is closely connected with their early nocturnal way of life. In order to survive in deep undergrowth or in the trees it is vital for an animal to have highly developed sense organs, particularly hearing, smell and touch. Mammals' sense of smell is very keen and they have three sound-conducting bones in the inner ear which give them a more acute sense of hearing than other animals. The three-bone structure acts as an amplification system and enables the mammal to pick up the faintest and most subtle sounds. Most mammals do not see particularly well but their sense of touch is well developed and many have specially sensitive hairs or whiskers on the face. Mammals therefore receive more information from their senses than other animals and the parts of the brain which process the information have grown larger to deal with it.

Memory, a capacity for learning and the ability to absorb a mass of information and choose a course of action in the light of experience have made mammals very adaptable. It is this which accounts for their enormous success and marks them off from the rest of the living world.

Who's who in the mammal world

The great variety of living mammals can be divided into three major groups, according to the way they reproduce.

The monotremes The monotremes are the most primitive group of living mammals and are now found only in Australasia and New Guinea. They are furry and feed their young on milk, but they lay eggs like reptiles and the adults have no teeth. There are two types, the echidnas or spiny anteaters and the duck-billed platypus. Female spiny anteaters develop a pouch at the beginning of the mating season and carry their egg there. They have no teats but milk oozes out from special milk glands and the young spiny anteaters lick it from the mother's hairs. The duck-billed platypus has no pouch, but builds a special den to protect her eggs. Like spiny anteaters, a young duck-billed platypus feeds by licking not sucking.

The marsupials The marsupials are pouched mammals. They are found in Australasia and in South, Central and southern North America. The young are born alive, but at a very early stage of development – an adult great grey kangaroo is over 2m high but is only about 2.5cm long at birth. The tiny animal makes its way to the mother's pouch and remains there, attached to the nipple, until it is sufficiently developed to feed independently. A few marsupials do not have pouches: instead their young cling to the mother's back.

There are four main groups of marsupials, the carnivores (flesh eaters), insectivores (insect eaters), bandicoots (which eat insects, small lizards and occasional roots) and the herbivores (plant eaters). The carnivores include American opossums, Australasian marsupial 'cats', 'wolves' and 'anteaters'. The insectivores are the rat opossums from South America and Australasian 'mice' and 'moles'. Bandicoots, from Australasia, are similar to African elephant shrews. The herbivores include phalangers, koalas and gliders, kangaroos and wallabies of various types and wombats.

The placental mammals Placental mammals are by far the most successful of the mammal groups and are found widely distributed all over the world, inhabiting all kinds of environments. In placental mammals the young remain inside the mother, nourished by food supplied through her bloodstream and protected from danger by her body. The placenta which gives the group their name links the blood systems of mother and baby so that food and waste products can be exchanged.

The length of time the young grow inside the mother varies, and so does their degree of development at birth. Antelopes, born on the open plains, are ready to run within minutes of being born while, at the other extreme,

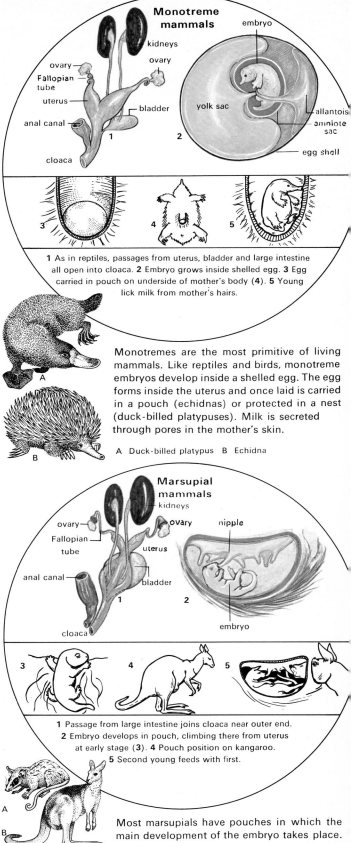

Monotreme mammals

embryo · kidneys · ovary · ovary · Fallopian tube · uterus · bladder · anal canal · cloaca · yolk sac · allantois · amniote sac · egg shell

1 As in reptiles, passages from uterus, bladder and large intestine all open into cloaca. **2** Embryo grows inside shelled egg. **3** Egg carried in pouch on underside of mother's body (**4**). **5** Young lick milk from mother's hairs.

Monotremes are the most primitive of living mammals. Like reptiles and birds, monotreme embryos develop inside a shelled egg. The egg forms inside the uterus and once laid is carried in a pouch (echidnas) or protected in a nest (duck-billed platypuses). Milk is secreted through pores in the mother's skin.

A Duck-billed platypus B Echidna

Marsupial mammals

kidneys · ovary · ovary · nipple · Fallopian tube · uterus · anal canal · bladder · cloaca · embryo

1 Passage from large intestine joins cloaca near outer end. **2** Embryo develops in pouch, climbing there from uterus at early stage (**3**). **4** Pouch position on kangaroo. **5** Second young feeds with first.

Most marsupials have pouches in which the main development of the embryo takes place. The mother helps her tiny young to crawl to the pouch, where it attaches itself to a nipple. When it leaves the pouch it continues to suckle but the milk changes. By this time a second young may be developing.

A Opossum B Kangaroo C Rat opossum D Bandicoot

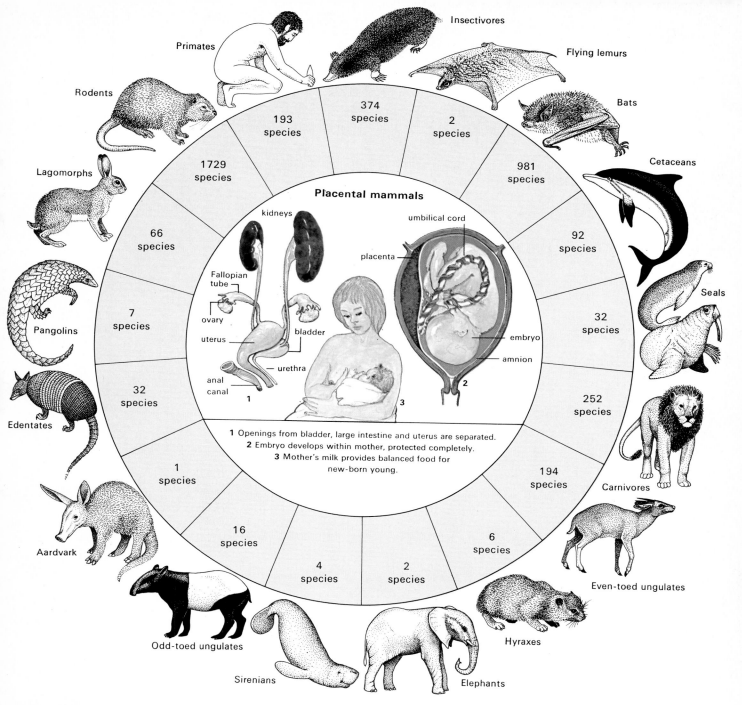

Primates

Insectivores

Flying lemurs

Rodents

Bats

Lagomorphs

Cetaceans

Placental mammals

Pangolins

Seals

kidneys

umbilical cord

placenta

Fallopian tube

ovary

bladder

uterus

embryo

urethra

amnion

anal canal

Edentates

Carnivores

1 Openings from bladder, large intestine and uterus are separated.
2 Embryo develops within mother, protected completely.
3 Mother's milk provides balanced food for new-born young.

Aardvark

Even-toed ungulates

Odd-toed ungulates

Hyraxes

Sirenians

Elephants

193 species

374 species

2 species

1729 species

981 species

66 species

92 species

7 species

32 species

32 species

252 species

1 species

194 species

16 species

6 species

4 species

2 species

Placental mammals are the most numerous and successful of all the mammals. They have evolved a special organ, the placenta, made up of tissue from both mother and embryo. Food and oxygen pass through this from mother to embryo; waste and carbon dioxide pass from embryo to mother. Placental mammals grow in the uterus for different lengths of time and are born at different stages of development. Some young can walk within minutes of being born, others are blind and helpless.

a human child may be a year or more old before he takes his first unsteady steps.

Because the chances of survival are relatively high, mammals give birth to only a few young at a time; some species always produce only a single offspring.

Living placental mammals are divided into 17 orders, and include 3,983 species. Some of the groups contain only a handful of species – one animal, the aardvark, is an order on its own. Of the seventeen groups, rodents are undoubtedly the most successful, accounting at the

moment for two-fifths of all mammal species.

The main evolution of the mammals has taken place over the last 64 million years, a relatively short time in the 4,600 million year history of the world. Though dramatic changes have taken place in some groups, other mammals have remained virtually unchanged. The insectivores (the shrews, hedgehogs and moles) are living members of the basic primitive stock which survived the 140 million year Age of the Dinosaurs and from which the great variety of mammals later evolved.

How Old are the Rocks?

The history of the mammals, like the history of all past life, is found in the rocks of the Earth's crust where the remains of animals and plants are preserved as fossils. As the surface of the land is gradually worn down (by rain, wind, ice and wave action) grains of sand, pebbles and mud are carried away by streams and rivers. When the rivers reach flat land and flow more slowly, the particles settle on the bottom and gradually build up deposits which harden into rocks. Animal bones carried by the rivers sink in the same places and are buried. Flood plains, estuaries, deltas and the shallow continental shelves around the edges of the land are places where deposits are most likely to be made.

As time passes, sediments accumulate and younger sediments are laid down in layers on top of the older ones. From the layers or succession of rocks or strata it is possible to see which rocks are older than others. Normally the older rocks and the fossils they contain are buried deep below the more recent strata. However, if earth movements lift the rocks, different layers and types will be exposed. No matter how much the rocks have been folded and contorted, it is still possible to work out which layers were laid down first.

Geological time is divided into different eras and periods, according to the most common type of fossil found in the different layers of rocks. Strata from places often far apart in the world can be given the same date because they contain the same type of fossil plants or animals. At first it was only possible to date rocks relatively, to say which one was older or younger.

The order, though not the time-span of geological history was established by the middle of the nineteenth century and by then it was also agreed that each of the major geological divisions had its own characteristic fossils. The fossils from each period were so distinctive that many geologists thought there had been a series of world-wide catastrophes, each one wiping out completely the forms of life that had existed before. Nobody believed in the idea that plants and animals had been gradually changing and evolving for millions of years.

Then, in the 1830s, Charles Lyell published a book which convinced scientists that even the oldest sediments had been formed by erosion from still more ancient rocks. The same processes they could observe in their own time had caused all the geological changes in the past. Charles Darwin applied the same argument to living things. He believed that by studying the way living animals varied from generation to generation, it was possible to work out how they had changed in the past. The only problem the two men could not solve was the length of time: vast ages of time were needed for

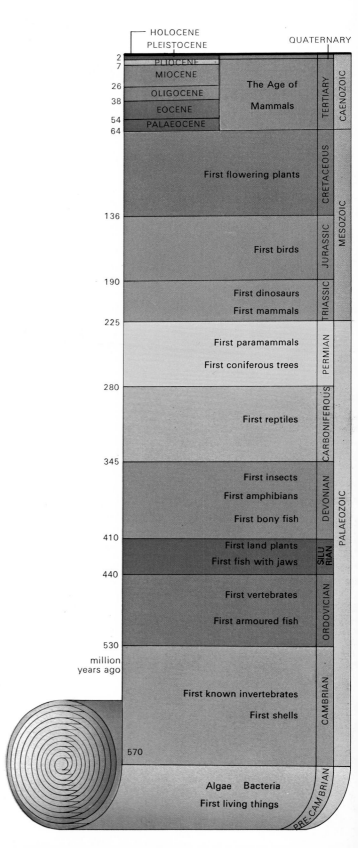

Geological time is divided into different periods which mark the important changes in animal and plant life recorded in the fossil record of the rocks. The Earth was formed about 4,600 million years ago but we know very little about the animals and plants that were evolving until around 600 million years ago, when large numbers of fossils are first found.

How sedimentary rocks are formed

Right: Rock strata in the Painted Desert, Arizona, contain many dinosaur fossils. Earth movements and weathering have brought rocks of different ages to the surface.

1 Wind and rain break down the surface of the land. Particles of sand, pebbles and clay are washed into rivers and carried along by the fast-flowing water.

2 When the river reaches flat land it flows more slowly and the particles or sediments drop to the bottom. In salt water clay particles group together to form mud.

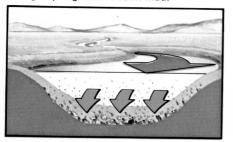

3 As the river flows on slowly across the plain, more and more sediments fall. They lie in layers or strata which are pressed together. As they harden, they form new rock.

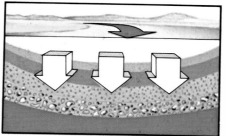

4 Later, the plain or sea bed may be raised by earth movements. New land is exposed and the cycle of erosion, weathering and deposition begins again.

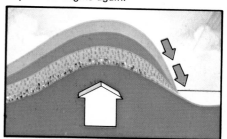

both the geological and the living world to have changed so much by the slow processes they could see at work and in those days the Earth was considered to be only 20 to 40 million years old. This was the length of time scientists calculated the Earth had taken to cool.

Then, at the end of the nineteenth century, radioactivity was discovered and it was proved that radioactive substances deep in the Earth's core prevented it from cooling down at the rate the earlier scientists had estimated. Everything could now be very much older than anyone had imagined; the time Darwin and Lyell needed for their theories was there.

Dating and geological time

Early in this century radioactivity began to be used to date rocks. Radioactive minerals occur in rocks from the Earth's inner core which have reached the surface, often erupting as volcanic lava. The radioactive element uranium 235 decays at a rate that can be measured and when it decays it changes into a special type of lead. By measuring the amount of lead and comparing it with the amount of uranium left in the rock, it is possible to calculate the length of time that has passed since the rock was formed. Other radioactive elements give even more accurate dates and using these, the length in years of all the geological periods has been carefully calculated.

The Earth itself is probably around 4,600 million years old: the oldest rocks actually found are nearly 4,000 million years old and are in the west of Greenland. For millions of years there is no fossil record of life of any kind, but by 3,200 million years ago the first simple one-celled plants and bacteria had evolved and by 1,900 million years ago, many-celled plants were living in the sea. It was not until animals began to develop hard shells some 600 million years ago that they were preserved in any numbers.

The thousands of millions of years between the formation of the Earth and the first major record of fossil animals and plants are called the Pre-Cambrian eras. The 570 million years since then are divided into three eras, the Palaeozoic, the era of early life; the Mesozoic, the era of middle life; and the Caenozoic, the era of recent life. Each era is subdivided into a number of periods and in the Caenozoic, where fossil remains are far more abundant, these are split into smaller epochs. 'Small' is rather a misleading description: even the shortest epoch lasted for 2 million years.

Mammals first evolved in the Triassic period, around 200 million years ago, but it was during the Caenozoic era, beginning 64 million years ago, that they became dominant. For this reason the Caenozoic era is often known as the Age of the Mammals.

Evolution and the Fossil Record

Evolution means literally 'unrolling' and describes the continuous development and change that has given us such a variety of plants and animals adapted for ways of life in different environments.

Over the thousands of millions of years since the Earth was formed, animals have developed from simple, single-celled creatures into more and more complex forms. Changes occur very gradually, often as animals adapt to changes in the world around them. Sometimes the environment does not alter and animals remain in their primitive form: blue-green algae very like those that first evolved 3,000 million years ago are still living.

An animal's place in nature is known as its ecological niche. This includes the kind of food it eats, the time of day it hunts or feeds, its particular size, the type of environment (desert, water, forest, plain) it prefers, in fact everything that concerns its special way of life. Unless food is extremely plentiful, only one type of animal can occupy a particular ecological niche at a time. Competition for the possible niches is an important influence in evolutionary change and you will see as you read this book that time after time different animals adapt to a new way of life *because* it becomes available. It is also very unusual for a possible niche to be left empty. If one type of animal ceases to use it, another will almost certainly become adapted to take its place.

Some evolutionary changes are the result of chance. An animal may develop a feature that gives it an advantage over other animals of the same species. It may, for instance, have a longer neck which enables it to eat the leaves of trees instead of plants growing on the ground. If food becomes scarce it will have a better chance of surviving because it will not be competing with the other animals for food: it will have found a new ecological niche. Its offspring will inherit its longer neck and will also have a better chance of living longer and breeding more long-necked young.

It is impossible to know the exact colours of most prehistoric mammals but these, too, evolved to suit the animals' particular way of life. Spots, stripes and camouflage colours that give protection in forests and grasslands all play their part in natural selection and ensure that the animals best suited to their environments survive.

Animals in an evolutionary line are called 'true' cats, dogs etc., when they are known to be genetically related. Their relationships can be traced from special features such as ridges in the teeth which prove one form is directly connected with another. Forerunners of the 'true' line may differ in some details but in others may be similar. 'Cat-like' or 'dog-like' animals are

animals in unrelated groups which fill the same ecological niche and look superficially the same.

Fossils: the proof of change

The story of evolution is a story of adaptation to new ways of life. Fossils, together with the study of living plants and animals, provide the evidence.

Only animals with hard parts such as bones, teeth, shells or skeletons, stand much chance of being preserved. The soft parts quickly rot away. Even the hard parts are soon broken down by weathering and bacteria unless they are rapidly covered by sediments.

Bone is made up of protein and mineral salts. The protein part gives the bones their shape while the mineral keeps them hard. When bones are buried in suitable sediments, the mineral of the bones and teeth is preserved and in some cases even the original protein can survive. In other cases the bone itself disappears but leaves an impression of its shape in the rock which surrounds it.

Of course many animals live and die in areas where there are no soft sediments to cover them and unless they are carried away by a river or flood to a more suitable

1 Fifty million years ago a bat, feeding on flying insects over a lake, swooped down too close to the water and was killed.

2 At first the gas in its guts made it float on its back with its wings spread out. Eventually it sank to the bottom of the lake.

site, they will vanish without trace. Most fossils of land mammals are broken bones that have been rolled along in rivers before being dropped into the sands and gravels of the river bed. In some places, vast heaps of bones have accumulated. In the 5 million year old deposits at Pikermi in Greece, for example, enormous quantities of bones from early three-toed horses *Hipparion* have been found.

With this type of fossil it is usually quite easy to see that it has been transported some distance before being buried but sometimes an animal may be carried away from its natural environment and still be perfectly preserved. About 50 million years ago the first known bat *Icaronycteris* floated dead on the surface waters of a lake. Dead bats float in water with their wings spread but when they sink to the bottom the wings fold up and they come to rest lying on their backs. As the soft parts of the bat rotted, the fine mud at the bottom of the lake covered its skeleton, preserving also fragments of wing membrane and diaphragm and even some of the food it had eaten at its last meal. In this case the structure of the skeleton proved that *Icaronycteris* was a flying mammal, even though it was found in sediments at the bottom of a lake, with perfectly preserved fish fossils. But it shows how important it is to consider not only where fossils are found but where they may originally have come from.

In places such as flood plains, deserts and coasts where there are moving sand dunes, deltas and estuaries, animals may be buried and fossilized in the place where they lived and died. When this happens it is often possible to discover a lot about the real environment in which they lived. Fossil plants preserved in the sediments around them and the nature of the sediments themselves show experts what the climate and vegetation was like while other fossil animals help to build up a picture of animal life millions of years ago.

One of the most interesting fossil mammal sites is the Rancho la Brea tarpits in Los Angeles. Here American mastodons, giant ground sloths, bison, horses, sabretooth cats and wolves were trapped in the sticky tar oozing out of the ground and their skeletons were buried and preserved together.

Sometimes animal fossils are found in caves. Ice Age cave bears, for example, often died during their winter hibernation while other animals fell down holes and

3 Fine mud particles were slowly accumulating on the lake bottom and these gently covered the bat.

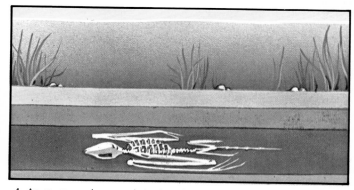

4 As more mud covered the bat, it was flattened. Over millions of years the mud hardened into rock with the fossil bat trapped inside.

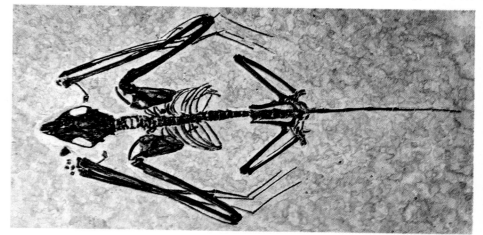

Right: Fifty million years after the bat died, earth movements brought rocks from the lake bed to the surface. When they were split open by fossil hunters, the fossil of the first known bat *Icaronycteris* was discovered.

fissures into underground caverns. The cave floors were rapidly covered with earth washed in by the rain and the bones remained undisturbed until they were excavated by fossil hunters.

The most spectacular 'fossils' are undoubtedly several complete animals that lived on land during the last Ice Age 100,000 years ago and have been perfectly preserved in the ice and snow. As early as 1692 a Dutch traveller in Siberia recorded the frozen remains of a mammoth, a species that had been extinct for thousands of years. Then, in 1900, a complete mammoth was discovered in the permafrost of the Yakutsk province of Siberia. The specimen was recovered and is on display in Leningrad's zoological museum. Unfortunately it is not complete for wolves and other wild animals had eaten the parts that had become exposed and thawed out before the expedition could save it.

In 1902 the remains of woolly rhinoceroses were discovered in a Polish mine and in 1929, again in Poland, a complete female woolly rhinoceros lying on her back was found. The animal's hair was not preserved but all her skin and internal organs were naturally pickled in oil and salt. In 1977 a complete, perfectly preserved baby mammoth was discovered in Russia.

Whole animals preserved in this way are very rare indeed. Fortunately a great deal can be discovered about the mammals of the past from their bones and teeth, the most common parts to be fossilized. From an animal's teeth it is possible to tell what kind of food it ate. If the teeth changed in some way it was usually because the species was adapting to a new kind of food and this in turn may mean that the environment in which it lived was changing. From just a few bone fragments experts can estimate an animal's size and deduce many other details about its structure and habits.

Fossils can also reveal something about animal behaviour. Bones dropped into rock fissures by leopards 2 million years ago show scratches and grooves where porcupines had gnawed them; a child's skull broken by a leopard's teeth and the partially healed head wound of an early cat provide direct information about violent incidents which actually happened. The bones of cave bears show that they suffered from severe osteoarthritis (a disease of the bones) as they grew old, while the skull of the 40,000 years old Rhodesian man proves that toothache is nothing new: he had large abscesses at the roots of his teeth, one so large that it had worn away the bone at the side of his head.

By putting all these aspects of fossil study together, it is possible to build up a realistic picture of prehistoric animal life millions of years ago.

Fossil skull of *Cynognathus*, a mammal-like reptile that lived over 200 million years ago. It was from animals like these that the first mammals evolved. The bones are now completely mineralized.

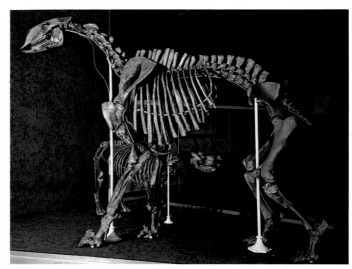

The skeleton of a chalicothere, a strange clawed animal that died out around 2 million years ago. The separate fossilized bones have been pieced together for display in a museum.

Thousands of years ago a baby mammoth fell into a crevasse and was deep frozen in the ice. It was discovered in Siberia in 1977, perfectly preserved with even its soft parts undamaged.

The Beginning of Life on Earth

The Earth is thought to have been formed about 4,600 million years ago. Rocks from the moon, which formed at the same time as the Earth but have not been eroded in the same way, have provided the date.

No-one knows exactly how or when the first life forms came about but we know that by 3,760 million years ago, some kind of primitive plants had evolved. No fossils have been found from this ancient time but some rocks contain a type of ironstone that cannot be formed without oxygen. Since at that time there was no free oxygen in the Earth's atmosphere, the oxygen needed to form the ironstone can only have come from early green plants, which use the energy of sunlight to manufacture food and produce oxygen as a by-product.

The first fossils of micro-organisms such as bacteria are found in 3,300 million year old rocks from Swaziland. By 2,900 million years ago there are fossils of blue-green algae, simple, single-celled plants which are still found today. There must have been countless numbers of plants by this time for by 1,300 million years ago they had produced enough oxygen for the first animals, burrowing and crawling water creatures, to evolve. By 680 million years ago there were a large variety of animals, all still living in the water where they were protected from the strong ultra-violet radiation reaching Earth at that time. The richest finds come from Ediacara in Australia where types of worms, sea-pens and jellyfish have been found as well as other animals that are unlike any living groups.

About 570 million years ago at the end of the long Precambrian era, several groups of animals began to develop hard skeletons to support their developing muscles. From this time onwards the fossil record becomes increasingly rich and the record of life can be traced in considerable detail.

The first vertebrates

Mammals belong to the group of animals known as Chordates, nearly all of which are vertebrates – animals with backbones. The first evidence of animals with backbones consists of fragments of bony armour from jawless fishes. They have been found in 500 million year old rocks from North America and Australia. These early fishes lived in the mud at the bottom of the sea and are believed to have evolved from the larvae of sea squirts or of rare animals called pterobranchs.

We know that 550 million years ago pterobranchs were living in the sediments at the bottom of the sea, filtering minute particles of food from the water. Though as adults they lived fixed in one place, their young or larval forms swam freely, just as the larvae of modern bottom-living sea squirts do today. Like modern larval sea squirts, the ancestors of the vertebrates would have been active feeders, swimming in the surface waters and eating the microscopic organisms that floated there. Also like the modern larvae, they would have had a muscular tail and a stiff rod of cells called a notochord, the forerunner of the backbone. Today, these larvae settle on the sea floor and the muscular tail and notochord, no longer needed for swimming, disappear.

If this were the only sea squirt life cycle it would be difficult to see how it could have led to animals with backbones. However, not all sea squirt larvae turn into bottom-living animals. Some change but remain at the surface while others never mature beyond their larval stage. It is now thought that millions of years ago some active larvae began to reproduce while they were still swimming freely and still had their notochord and tail.

As these tiny animals fed, they absorbed calcium salts, which they stored in their skin as calcium phosphate. Phosphates are an important part of animal diet and

The first vertebrates

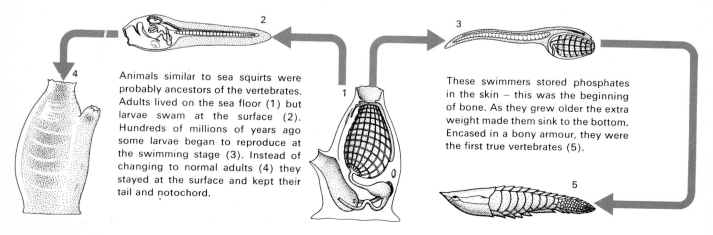

Animals similar to sea squirts were probably ancestors of the vertebrates. Adults lived on the sea floor (1) but larvae swam at the surface (2). Hundreds of millions of years ago some larvae began to reproduce at the swimming stage (3). Instead of changing to normal adults (4) they stayed at the surface and kept their tail and notochord.

These swimmers stored phosphates in the skin – this was the beginning of bone. As they grew older the extra weight made them sink to the bottom. Encased in a bony armour, they were the first true vertebrates (5).

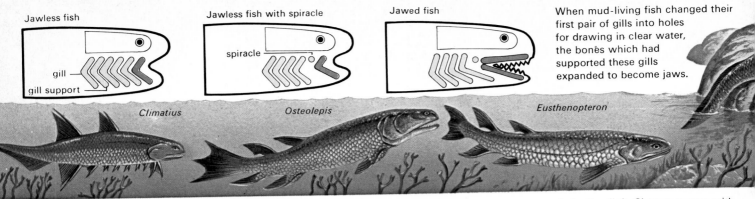

Jawless fish

gill

gill support

Jawless fish with spiracle

spiracle

Jawed fish

When mud-living fish changed their first pair of gills into holes for drawing in clear water, the bones which had supported these gills expanded to become jaws.

Climatius

Osteolepis

Eusthenopteron

The first jawed fishes had stiff, triangular fins (A) which acted as stabilizers to help them control their position in the water. More manoeuvrable fish developed in two ways, as ray-finned (B) and lobe-finned (C) types. Ray fins were narrow near the fish's body but then fanned out into a wide 'ray' shape. They did not evolve into limbs.

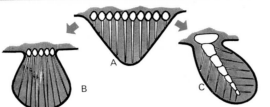

Lobe fins (left, C) were narrow with strong muscles and a row of central bones. The bones of the fins were arranged in exactly the same way as the bones of the legs of all land vertebrates (right). All the different types of vertebrate limbs can be traced back to lobe fins.

animals that were able to store them would have had a great advantage in seasons when they were not easy to obtain. In time the calcium phosphates hardened into a protective covering. As this increased the animals' density, they could no longer live on the surface but sank to the bottom and lived in the mud.

It may never be possible to know exactly what these early vertebrate ancestors looked like and how quickly the notochord strengthened into a true backbone. But 50 million years later they had evolved into true vertebrates, the jawless armoured fishes.

The evolution of the fishes

One of the major events in the evolution of the vertebrates was heralded some 400 million years ago when plants and water animals invaded freshwater lakes and rivers. The animals included some of the jawless mud-living fishes. However, the next important structural change took place in animals that remained in the sea. Some of these became specially adapted for living in muddy conditions. One pair of their gills was converted into a passage which opened on the top surface of the head, behind the eyes. Water could be drawn in through this without mud getting in the gills. Because the first pair of gills had disappeared, there was room for the skeleton in that part of the animal's body to expand. The bony top and bottom parts of the mouth grew stronger to form the very beginnings of jaws. The vertebrates could now begin a new way of life as predators and their jaws soon became sufficiently powerful to bite pieces out of their fellow animals. They, too, gradually moved into freshwater environments.

The first jawless vertebrates did not have flexible fins to control their movement. The front part of their body was wide and gave them a basic stability and the tail, flattened from side to side, gave them a forward thrust. By moving the tail they could lift themselves out of the

sand or mud but as soon as it stopped its side to side swing, the animal glided back to the bottom. When predators evolved, both they and their prey had to develop a more efficient way of swimming – the predators in order to hunt better, the prey to be able to escape.

The earliest jawed or bony fishes had pairs of triangular fins down their body, developed from simple stabilizing outgrowths. In time these were reduced to just two main pairs, the pectorals at the front and the pelvics further back.

These fishes could now actually control their movement through the water and gradually the fins changed from their triangular shape and became even more mobile. In one group of fishes, known as the ray-finned fishes, the bones inside the fins became closely grouped at the base of the fin, then spread out into a ray. In the second group, the lobe-finned fishes, the central fin bones grew stronger and the outer ones disappeared, so that the fins became oval or lobe-shaped.

These changes occurred in freshwater bony fishes. The same thing happened quite independently in the sea where early sharks and shark-like fishes were living. The earliest sharks had broad based triangular fins but later evolved into ray-finned (e.g. skates, rays and sharks) and lobe-finned types (extinct sharks).

The Age of Fishes

During the Devonian period (410-345 million years ago) the fishes evolved into so many different groups that the period is known as the Age of Fishes. They were successful in rivers and lakes in many parts of the world as well as in the oceans. But the semi-tropical climate, with its alternating droughts and rainy seasons, created serious problems for all freshwater plants and animals. When rivers dried up the fishes could only survive in ponds and lakes. The oxygen dissolved in these waters was

Ichthyostega

Gephyrostegus

Hylonomus

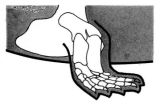

Although it still had a fish's tail and lived most of its life in the water, *Ichthyostega* was the first amphibian. It had good lungs for breathing air and sturdy limbs. Later amphibians such as *Gephyrostegus* were even better adapted for land life but still returned to the water to breed.

All higher vertebrates are descended from romerids like *Hylonomus*, a small lizard-like reptile that lived some 300 million years ago. The reptiles' shelled egg (left) was laid on land. It enclosed the embryo in a private, watery pond and supplied it with food (yolk) and air.

rapidly used up and the fishes died of asphyxiation.

By the Devonian, plants had produced enough oxygen to create a barrier against ultra-violet radiation and were well established on land. There was also now a large proportion of oxygen in the Earth's atmosphere. Most of the later Devonian fishes developed some kind of extra breathing organ so that they could use oxygen directly from the air as well as through their gills under water. In this way they were able to survive the most difficult time of year.

One group of fishes, the lobe-finned fishes, had a further advantage. Their fins had a strong central group of bones and they could use them like legs to crawl short distances on land. If the pond in which they were trapped dried up, they could drag themselves with their strong muscular fins across mud or sand to another stretch of water. During the annual drought these four-legged fishes had an advantage over all the other water animals and their chances of survival were greatly increased. Animals such as the 350 million year old *Ichthyostega* were the first amphibians. They could move about either in the water or on land – even if the land, as far as they were concerned, was only a last resort.

The Age of Amphibians

During the next period, the Carboniferous, lush tropical forests covered large parts of the continents and vast numbers of all kinds of small water animals lived in the swamps. At this time the four-legged fish, now true amphibians, came into their own and an enormous variety evolved. Most became adapted to life in shallow water. They became flatter and their limbs became weaker. One group kept its deep fish-like body and strong limbs and was able to hunt insects and other small animals in the steamy tropical forests.

One of the main problems for amphibians is their need to lay their eggs in water for their young tadpoles

breathe like fish, through gills. Young amphibians are extremely vulnerable and the adults lay thousands of eggs so that at least a few will manage to survive to keep the species alive. However, if the eggs are hidden they have a better chance of hatching, so the mother can lay fewer. This is more efficient and avoids the enormous waste of producing a mass of eggs, most of which will simply provide food for other animals. Tropical frogs of today often lay their eggs away from water: some hide them under logs, others such as the Surinam toad *Pipa* actually carry them in pockets in the skin of the mother's back until they have developed into young adults.

During the Carboniferous period one group of amphibians began to lay its eggs on land – by far the safest place, for most animals were still living in the water. The eggs had a large yolk to provide food for the developing embryo. This was surrounded by a membrane, the amnion, which enclosed a private, watery pond. A further membrane, the allantois, acted as a lung and the whole egg was enclosed in a protective covering, the chorion. A chalky shell, porous enough to allow oxygen to diffuse in and carbon dioxide out, was an extra protection and also provided calcium to form the embryo's bones. This type of egg, the amniote or cleidoic (meaning closed) egg, gave these animals a big advantage in the Carboniferous swamps. The adults no longer needed to return to the water to lay their eggs but could live permanently in the new land environment where there was less competition for food.

Animals that produce this type of egg are known as reptiles and when the swamps dried up at the end of the Carboniferous 280 million years ago and most of the amphibians became extinct, the reptiles survived and gradually spread over the continents. They included a group called the paramammals (or mammal-like reptiles), from which, millions of years later, the mammals themselves eventually emerged.

Life on the Land: the Paramammals

The first reptiles to flourish in the coal forest swamps of the Carboniferous period were small, lizard-like creatures called romerids. They were agile enough to scamper over fallen tree trunks in their hunt for insects and their remains have been found in the hollow stumps where they sometimes were trapped. From this primitive stock two major evolutionary lines of reptiles evolved: one, known from a fossil called *Petrolacosaurus*, was the ancestor of the dinosaurs and the birds. The other group was the primitive paramammals, the pelycosaurs which eventually gave rise to the mammals.

Towards the end of the Carboniferous, the lush swamps dried up and there was generally a drier climate everywhere. The reptiles, which had already adapted to some extent to life on land, were now in a strong position. For the time, the descendants of *Petrolacosaurus* remained small insect-eaters but the other group, the paramammals, began an important evolutionary expansion.

The new environment brought great changes to the paramammals' way of life although at first they continued to feed in and around the remaining rivers and lakes. The move to a life wholly on the land was not a simple one. At this time none of the vertebrates had digestive systems that could cope with plant food; they had to feed on other animals that had eaten plants or on one another. The smallest paramammals ate invertebrates such as insects, worms and molluscs; the medium-sized ones ate the smaller paramammals and the largest ate the medium-sized ones and one another.

The primitive paramammals had long snouts with a large number of small, sharp teeth. These formed what was essentially a fish trap to prevent prey from slipping out into the water. The small teeth were too weak to hold on to struggling reptiles on the land, and gradually the paramammals changed to cope with their new way of life. Their jaws became shorter and more powerful and although they had fewer teeth, some of the front ones grew longer to act as stabbing daggers. They were now able to hold their prey firmly and kill it quickly. Their skull also became stronger and inside their mouth they developed a soft palate which separated the air passage to the lungs from the food passage to the stomach. This meant that the animal was able to breathe continuously even when it was feeding, and had a constant supply of oxygen.

Dimetrodon

Edaphosaurus

Ophiacodon

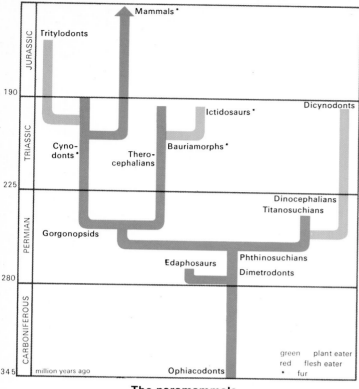

The paramammals

The dominant land animals in the Permian were the flesh-eating paramammals. All were cold blooded. The larger types such as *Dimetrodon* and the shellfish eater *Edaphosaurus* controlled their temperature with heat absorbing 'solar panels'. Others such as *Ophiacodon* were small enough to warm up quickly in the usual way. *Petrolacosaurus* and its descendants fed on insects and worms and were the least important vertebrates of the time. Yet it was from these small lizard-like reptiles that the dinosaurs and birds evolved.

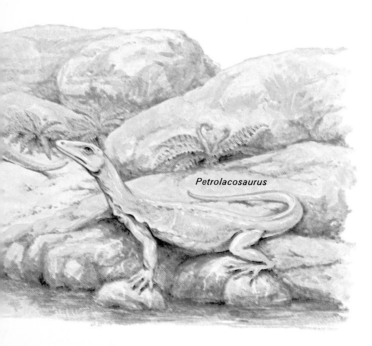

Petrolacosaurus

In true mammals this led to warm-bloodedness but the paramammals controlled their temperature in a different and in the end a less successful way. Some grew solar heating panels, sheets of skin supported by the backbone to make a large sail. When they turned the sail to face the sun the blood vessels inside it were heated and the warmth was soon carried round the whole body. In quite a short time the early morning sun's rays would warm the animal enough to make it active.

Dimetrodon, one of the most advanced of the early paramammals, was about 3.5m long and weighed 250kg. With its sail it could raise its temperature 6°C in 50 minutes, whereas without a sail it would have taken between 100 and 200 minutes. The average *Dimetrodon* took about 64 minutes to warm up in the mornings. Smaller paramammals such as *Ophiacodon* (1–1.5m long, 30–50kg) do not seem to have had sails but they were small enough to warm up more quickly and be on the move before, or at least at the same time as the larger carnivores.

Advanced paramammals

By the middle of the Permian, about 50 million years after the primitive paramammals had evolved, a more advanced type appeared, the theraspids. Until recently there was a gap in the fossil record between the primitive and advanced animals but now a fossil known as *Phthinosuchus* has been found in Perm, west of the Ural mountains, and this forms a link between the two groups.

The most important change was that plant eaters developed – massive, heavily built animals with dome-shaped heads. Because they were so large they were less vulnerable to the flesh eaters but the most significant thing about them was that they could feed directly on plants. The first plant eaters were called deinocephalians and from these another group, the dicynodonts, developed. These were to become the most successful of all the paramammals and dicynodonts of all sizes spread throughout the world. Their fossil remains have been found in rocks from the Antarctic, India and Africa.

Plant-eating vertebrates are important because they can digest plant material directly and do not have to rely on invertebrates such as worms and insects to process it for them. There is always more plant than animal food available and once vertebrates had learned how to make use of it, they were able to increase in numbers and also to become more varied. Some dicynodonts were large and heavy like the deinocephalians; others lived like hippopotamuses, often in the water; still others were specialized for digging underground roots.

The flesh-eating paramammals also developed several different evolutionary lines and their jaws and teeth became even stronger and more efficient. Both flesh and plant eaters also began to change the way they stood and walked, so that they could move more swiftly. The early paramammals walked with a sprawling gait, their upper limbs sticking out sideways from their body. The later ones had straighter legs and their body was higher off the ground.

Although they could now run faster, they were less well balanced than before, when their upper limbs had provided a kind of sling to support the body. To correct this, the cerebellum, the part of the brain concerned with muscular co-ordination and balance, increased. At the same time their neck became more flexible so that although the head was held higher off the ground, it could be lowered for feeding.

The first mammals

The story of the paramammals shows a series of gradual steps leading from an animal that is obviously a reptile to one that is in many ways like a mammal. Two groups of paramammal carnivores, the cynodonts and the bauriamorphs, were even warm blooded so that they could hunt actively regardless of the outside temperature. They had bony secondary palates separating their food and air passages and they had cheek teeth specialized for chewing. They may also have had a furry or hairy coat for there are small pits on fossil jaws where whiskers may once have grown. Anyone seeing one of these paramammals alive today would instinctively recognize it as a primitive mammal.

When, then, did the first real mammals evolve and how are they distinguished from these advanced paramammals? Scientists have chosen a change in the jaw joint as the dividing line. In reptiles (including paramammals) the jaw joint is hinged on two small bones, one linked to the skull, the other to the lower jaw. In mammals, the lower jaw bone itself is hinged directly to the skull. The small bones that form the reptile jaw joint are incorporated into the mammal's middle ear, where they act as sound conductors to give mammals their acute sense of hearing.

Moschops

Dinodontosaurus

Titanosuchus

Moschops

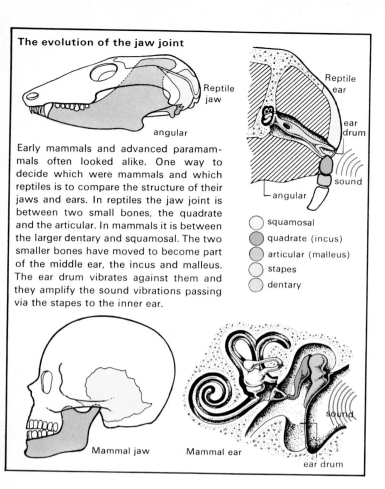

The evolution of the jaw joint

Early mammals and advanced paramammals often looked alike. One way to decide which were mammals and which reptiles is to compare the structure of their jaws and ears. In reptiles the jaw joint is between two small bones, the quadrate and the articular. In mammals it is between the larger dentary and squamosal. The two smaller bones have moved to become part of the middle ear, the incus and malleus. The ear drum vibrates against them and they amplify the sound vibrations passing via the stapes to the inner ear.

○ squamosal
● quadrate (incus)
● articular (malleus)
○ stapes
● dentary

This change from one type of jaw joint to another was so extraordinary that scientists believed it could only have happened in one group of animals, from which all mammals were descended. However, it is now known that several different types of paramammals were evolving mammal-like jaw joints, and probably more than one group crossed the reptile-mammal boundary.

Towards the end of the paramammals' history a group of small carnivores developed and from these, 200 million years ago, the first true, warm-blooded, suckling mammals evolved.

The advanced paramammals, the theraspids, evolved some 50 million years after the early flesh eaters. Giant lumbering plant eaters such as *Moschops* were forerunners of dicynodonts such as *Lystrosaurus* and *Dinodontosaurus*. Fierce flesh eaters such as *Titanosuchus* were ancestors of both the cynodonts (*Thrinaxodon* and *Cynognathus*) and the bauriamorphs (*Bauria*). Furry, warm-blooded flesh eaters, these were only a step away from being mammals. *Morganucodon*, the first true mammal, evolved around 200 million years ago.

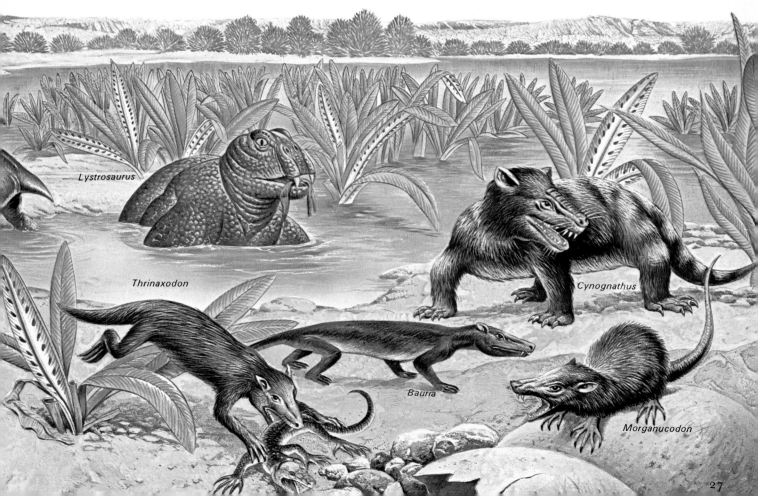

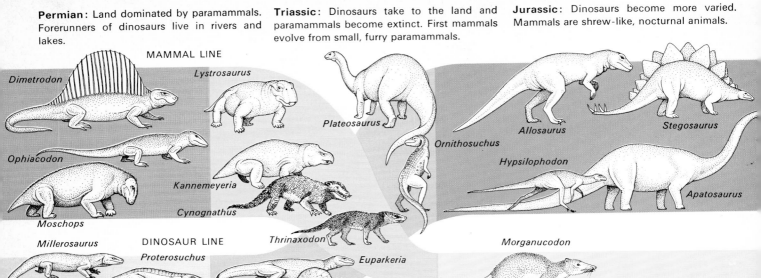

Permian: Land dominated by paramammals. Forerunners of dinosaurs live in rivers and lakes.

Triassic: Dinosaurs take to the land and paramammals become extinct. First mammals evolve from small, furry paramammals.

Jurassic: Dinosaurs become more varied. Mammals are shrew-like, nocturnal animals.

MAMMAL LINE

Dimetrodon

Lystrosaurus

Plateosaurus

Allosaurus

Stegosaurus

Ophiacodon

Ornithosuchus

Hypsilophodon

Kannemeyeria

Cynognathus

Apatosaurus

Moschops

Thrinaxodon

Morganucodon

Millerosaurus

DINOSAUR LINE

Proterosuchus

Euparkeria

Tortoise

The Age of Dinosaurs

During the time when the paramammals were dominant, the other reptiles – ancestors of lizards, crocodiles and dinosaurs – had been small insect eaters. About 250 million years ago the lizards began to develop into different forms. There were fish-eaters with enormously long necks, general beachcombers, shellfish eaters, even gliders. The dinosaurs and crocodiles evolved from a group of insect-eating reptiles called archosaurs which came to live like modern crocodiles, hunting and scavenging along river banks. It was a very different life from that of the parammamals, and while they remained mainly water animals, they did not compete.

As time passed, the archosaurs grew powerful tails and their hind limbs became longer and stronger than the front ones. This made them more efficient swimmers but also affected the way they walked on land. Their sprawling walk changed to a straight-limbed, longer stride but, unlike paramammals, they could not run fast on four legs because of the difference in lengths of their front and back legs. Instead they ran on their hind legs only, the heavy, muscular tail acting as a counterbalance to the head and body.

These active, fast-running carnivores seem to have taken to the land quite suddenly and within a few million years the new race of dinosaurs had almost completely destroyed both plant- and flesh-eating paramammals. The only paramammals to survive were the small rodent-like animals that were gradually evolving into true mammals.

When the paramammals disappeared, the world was once again inhabited by flesh-eating reptiles. However, from these, new plant eaters developed to take the place of the plant-eating paramammals and in time the dinosaurs produced a much greater variety of forms. Some of the small insect-eating reptiles developed into

parachutists and others into warm-blooded, furry pterosaurs which were true flyers. Another group evolved feathers which helped them to fly and also acted as insulation. From these feathered dinosaurs or first birds (such as *Archaeopteryx*) modern birds were later descended.

At this time the early mammals were small insect eaters. The remains of many of them have been found in cave deposits in South Wales. Other caves nearby contain the remains of small lizards, also insect eaters. Although the two groups of animals were competing for the same type of food, they hunted at different times and were therefore able to exist side by side. The lizards were active during the day, the mammals, being furry and warm blooded, could hunt at night when the temperature cooled the lizards into a torpor.

At the end of the Triassic, 190 million years ago, there were two major groups of mammals. The first is known from fossils named *Morganucodon* and *Megazostrodon*. A complete skeleton of *Megazostrodon* was found in Lesotho and shows that these animals may be distantly related to modern monotremes – the echidnas and platypuses.

The second major group is known from a fossil named *Kuehneotherium* which was found in South Wales. Animals of this type are thought to be the ancestors of both marsupial and placental mammals. *Kuehneotherium* was a small, shrew-like animal and probably looked outwardly much like the other mammals of the time. However, there are important differences in its teeth, which are very similar in basic pattern to modern mammal teeth.

Later, during the Jurassic period (190–136 million years ago) another group of mammals evolved. These are called the multituberculates because their teeth had tubercules or small lumps on the surface. These rat-like animals gnawed and nibbled their food and took the

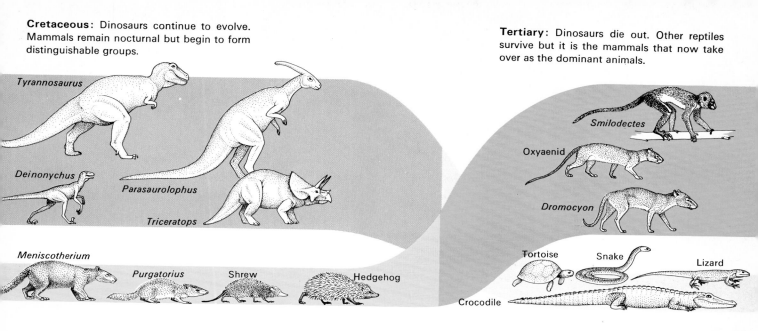

Cretaceous: Dinosaurs continue to evolve. Mammals remain nocturnal but begin to form distinguishable groups.

Tyrannosaurus

Deinonychus

Parasaurolophus

Triceratops

Meniscotherium

Purgatorius

Shrew

Hedgehog

Tertiary: Dinosaurs die out. Other reptiles survive but it is the mammals that now take over as the dominant animals.

Smilodectes

Oxyaenid

Dromocyon

Tortoise

Snake

Lizard

Crocodile

place of the small gnawing paramammals when they at last became extinct. The multituberculates lived very successfully for almost a hundred million years – much longer than the whole Age of Mammals. Like all early fossil mammals, their remains are rare but they have been found in every continent except Antarctica, Australia and South America. Eventually they died out completely and they are not the ancestors of any modern mammals.

Marsupial and placental mammals

In the middle of the Cretaceous period mammals divided into two distinct groups, the marsupials which carry their developing young in pouches and the placentals, which retain the young inside the mother's body. The earliest marsupials were primitive opossums, ancestors of today's North American opossums. The first placental mammals were similar to shrews and hedgehogs. Early monotremes continued unchanged.

By the end of the Cretaceous, the ancestors of more placental mammal groups were beginning to appear. The first primate, *Purgatorius*, similar to living tree shrews, was alive in North America. Ancestors of cat-like flesh eaters and scavengers have been found in North America and Mongolia. Finally, an important group called condylarths evolved. These were the ancestors of dog-like flesh eaters and of many types of hoofed plant eaters.

Though it is possible to separate these early mammals into distinct groups, they must have looked very like one another and lived in much the same kind of way. Most still fed on insects and other small animals, though some of the condylarths and the rat-like multituberculates also ate plants. At the time they evolved the other available niches were well filled by the many different

types of dinosaurs. Birds, too, were beginning to develop. There were water birds such as rails and waders, cormorants and flamingoes; one group of birds had even become adapted to prey on the small nocturnal mammals: the first owls had evolved.

The end of the dinosaurs

At the end of the Cretaceous period, about 64 million years ago, the dinosaurs and pterosaurs which had dominated the world for 140 million years became extinct. This was perhaps the most dramatic event in the history of life on Earth. How did it come about?

Certainly changes in climate were a major factor. Hot seasons were becoming hotter at this time and cold seasons colder and the larger dinosaurs, some weighing as much as 30 tonnes, would not have been able to remain active in cold weather. Nevertheless we now think that the final extinction of the dinosaurs was due largely to the mammals. When competition for food increases, small animals distributed over a wide area stand a better chance of survival than larger ones. A single duck-billed dinosaur weighing 3 tonnes was equivalent to 100,000 shrew-like mammals or 10,000 rat-sized ones; a 30 tonne brontosaur was equivalent to a million shrews and needed a corresponding amount of food. If there was a plague of mammals dinosaurs would have found it hard to obtain enough plants to eat.

Recent discoveries in North America show that towards the end of the Cretaceous period the dinosaurs retreated away from the area in which the mammals were most successful. As they retreated, the mammals followed, devouring the available food as they went. The dinosaurs were quite literally starved to extinction. The Age of the Dinosaurs had come to an end. The Age of Mammals was about to dawn.

29

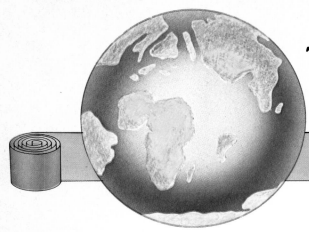

The Age of Mammals begins: the Palaeocene Epoch

Two hundred million years ago, at the time when the paramammals first evolved into true mammals, the Earth's surface looked very different from the way it does today. All the continents were joined together in a single land mass named Pangea. It had a southern and a northern arm, with in between the great Ocean of Tethys. Because the land mass was not divided and the climate was basically similar everywhere, the animals could spread over very wide areas. The same types of early mammal have been found in western Europe, China and southern Africa. They almost certainly also flourished in other areas where conditions were not suitable for their fossils to be preserved.

During the later Jurassic, Pangea began to break up and by the early Cretaceous, 130 million years ago, Antarctica and Australia were separated from the other southern continents and formed a single land mass. Africa and South America were linked only where Nigeria and eastern Brazil now are.

By the time Australasia and Antarctica broke away to form a separate unit the first marsupials had appeared, but there were not yet any placental mammals. This gave the Australian marsupials a tremendous advantage: they were able to develop into many different forms to fill all the available ecological niches, without having to compete with more adaptable animals.

By the end of the Cretaceous the Atlantic had widened and South America and Africa were no longer joined together. Europe and North America were still connected by land and South America was still linked by a narrow bridge to North America. However, at the beginning of the Age of Mammals, 64 million years ago, South America became totally isolated from all the other continents. The few mammals that had already evolved while it was joined developed quite separately from mammals on the other continents. Africa also became separated from the rest of the continents at about this time and here, too, the mammals developed in isolation.

North America, Europe and Asia remained joined and it is from the basic stock of placental mammals on these three continents that today's mammals developed.

After the dinosaurs

When the all-dominant dinosaurs disappeared from the Earth, the small nocturnal mammals were left in an essentially empty world. They no longer had to be active only at night but could venture out in daylight without danger. All the ecological niches formerly occupied by the dinosaurs – which had ranged from giant plant eaters to small, active insect eaters – were suddenly free. In spite of this, there seem to have been no dramatic developments to take advantage of the new opportunities. For several millions of years most mam-

Pantodonts lived between 64 and 38 million years ago. The first, *Pantolambda*, was about the size of a sheep and spent most of its time in water. *Barylambda* evolved later in the Palaeocene and lived entirely on land. Early condylarths such as *Meniscotherium* were about the size of a small dog. They were the basic ancestral stock from which most of the later plant eaters were to descend.

Pantolambda

Meniscotherium

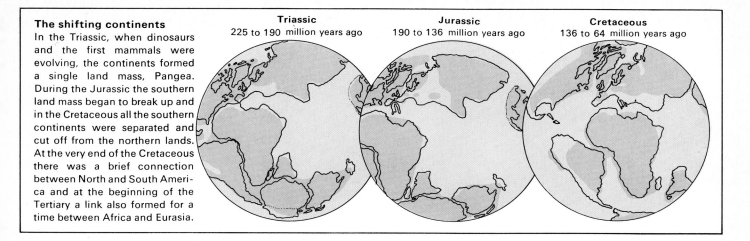

The shifting continents
In the Triassic, when dinosaurs and the first mammals were evolving, the continents formed a single land mass, Pangea. During the Jurassic the southern land mass began to break up and in the Cretaceous all the southern continents were separated and cut off from the northern lands. At the very end of the Cretaceous there was a brief connection between North and South America and at the beginning of the Tertiary a link also formed for a time between Africa and Eurasia.

Triassic
225 to 190 million years ago

Jurassic
190 to 136 million years ago

Cretaceous
136 to 64 million years ago

mals continued to live in trees or in deep undergrowth.

Change was almost imperceptible. Shrews still scurried about the undergrowth. Primitive hedgehogs came out at night to feed on worms and insects. Their way of life did not change – nor indeed has it altered in the millions of years since. They are perfect examples of groups which, having found a suitable ecological niche, have no reason to change. Marsupial opossums, too, continued as before. These opossums were very adaptable: they could eat virtually anything smaller than themselves, whether animals or plants. Even if one type of food was no longer available, something else would be just as good.

The tree-living primates, however, did make some advances. They filled the niche now occupied by rodents, gnawing and nibbling hard plant food. Like later primates, they had fingers and toes adapted for grasping things.

In both South and North America the first armadillo-like anteaters appeared. In North America these died out but in the south they flourished and became the ancestors of sloths, anteaters and armoured armadillos.

Among the other small mammals, the multituberculates continued to thrive and animals like squirrels and marmots appeared. The first lagomorph (the

Barylambda

Next page: Mammals of the Palaeocene forests. Tree livers included early primates (*Plesiadapis* 1) and gliders (*Planetotherium* 2). In the dense undergrowth were insect-eating shrews (*Palaeoryctes* 3) and hedgehogs (*Prodiacodon* 4), hunted by miacids (*Protictis* 5). Rat-like multituberculates (*Taeniolabis* 6) survived from the Jurassic, when they had co-existed with the dinosaurs.

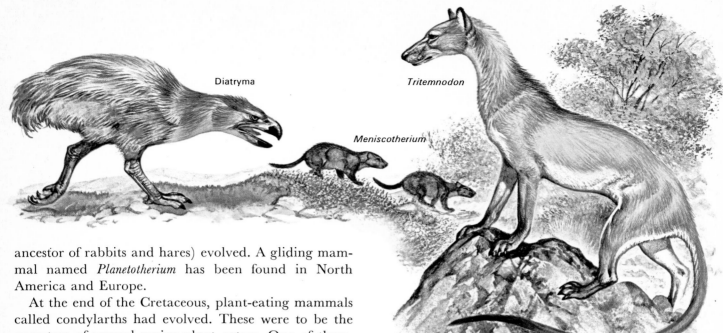

Diatryma

Tritemnodon

Meniscotherium

Sixty-four million years ago the most important flesh eaters were large flightless birds (diatrymas) and small long-bodied hyaenodonts such as *Tritemnodon*. They hunted the small plant eaters which lived in the forest undergrowth – animals such as *Meniscotherium*.

ancestor of rabbits and hares) evolved. A gliding mammal named *Planetotherium* has been found in North America and Europe.

At the end of the Cretaceous, plant-eating mammals called condylarths had evolved. These were to be the ancestors of several major plant eaters. One of them, *Meniscotherium*, is probably typical of the group. Little larger than a fox terrier, it had claws on its toes and a heavy tail. In some ways its teeth and skeleton are like those of modern African hyraxes and it may have been their distant ancestor.

An environment new to true mammals was taken over by heavily built animals with short, stocky limbs called pantodonts. One of the first, *Pantolambda*, was about the size of a large sheep and is known to have lived in both Asia and North America at this time. It was a plant eater, living as hippopotamuses do today – spending much of its time wallowing in the water.

Towards the end of the Palaeocene another pantodont, a strange-looking animal named *Barylambda*, appeared. It was about 2.5m long with thick, strong legs and a short, heavy tail. It is difficult to imagine how this oddly proportioned animal lived. By resting on its haunches and supporting itself with its tail, it could have reached for branches higher up the tree than other ground-living animals and this may have given it an advantage for a time. It survived for only a few million years – a short time in evolutionary terms. By early in the next epoch, the Eocene, it had become extinct.

The largest animals were the uintatheres which evolved towards the end of the Paleocene epoch, about 54 million years ago. The first were probably little bigger than modern tapirs but were heavily built and had several bony knobs on their heads to protect them from attackers. They were the equivalent, but not the ancestors, of modern rhinoceroses.

The flesh eaters

Several different types of flesh eaters evolved at around the same time as the plant eaters. All developed teeth specialized for cutting and slicing flesh but in other ways they were different from each other and hunted different types and sizes of prey. These carnivores were all primitive animals and most are not directly related to modern forms. Nevertheless it is possible to compare them to living animals.

There were five main groups: a cat-like group called oxyaenids, mongoose-like hyaenodonts and small, lightly built miacids were descended from the early insect eaters. The dog-like mesonychids and heavy, bear-like arctocyonids were descended from the condylarths. In spite of the evolution of all these different flesh eaters, the most important predators in the Palaeocene were not mammals at all. Along the edges of the rivers and lakes, crocodiles were the main danger while on dry land, giant flightless birds, the diatrymas, preyed on mammals of all sizes.

The end of the Palaeocene

By the end of the Palaeocene several important mammal lines had begun to evolve and to specialize for the different ecological niches left empty by the dinosaurs. Competition in the forests which still covered the lowland areas of most of the land was increasing and animals which could adapt to new ways of life were likely to be most successful. Not all the animals that evolved in the ten million years after the dinosaurs disappeared were to survive but some, as we shall see, can be linked more or less directly with modern species.

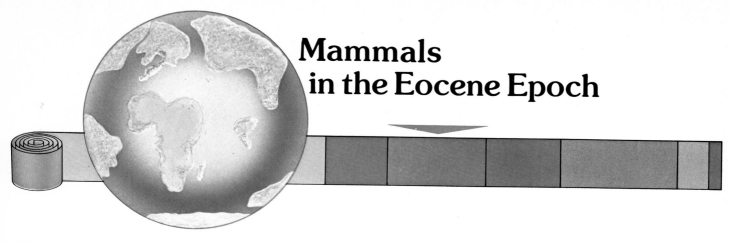

Mammals in the Eocene Epoch

During the next epoch, the Eocene, mammals became firmly established in many more environments. They dominated the land; they took to the air and they returned to the sea.

At the beginning of the Eocene, 54 million years ago, North America and Europe were connected by land and several animals were common to both places. There was also a land link between North America and Eastern Asia, across the Bering Straits. Asia, however, was separated from Europe by a shallow sea. The southern continents of South America, Africa, India and Australia remained completely cut off from the northern land masses as well as from one another, and mammals there continued to evolve in isolation. The mainstream of evolution was still in North America.

The climate of the Eocene was much less varied over the world than it is today. Even in the latitude where London lies, it was semi-tropical. Cycads and palms flourished, with figs, magnolia and cinnamon. The lowlands were densely wooded with many modern tree species and birds, including ducks, vultures and songbirds, were there. Modern lizards and snakes survived from the age of the dinosaurs and crocodiles, turtles and terrapins lived on the sea shores as well as in the lakes and rivers.

At first most mammals continued to live in the trees and to be active at night. More rodent-like primates had evolved but these were soon driven out by true rodents such as the squirrel-like *Paramys* and by the dormice. True rodents have chisel-like incisor teeth which grow continuously throughout their life. They are ideal for gnawing and nibbling tough plant material, for they never wear away. Only the less specialized early primates which could feed on a variety of food from fruit and nuts to small birds and eggs, survived this competition. One of these, *Notharctus*, had eyes placed towards the front of its face making it possible for it to judge distances accurately as it jumped from tree to tree. Another, *Tetonius*, had very large eyes – an adaptation for a nocturnal way of life. *Tetonius* may be an ancestor of the tarsiers.

Other small mammals such as shrews and hedgehogs lived among the undergrowth and on the forest floor and the first pocket gophers evolved. By the end of the Eocene the first true rabbits and hares had appeared.

Flying mammals

The most important change among the smaller mammals was that they learned to fly. The earliest known bat, *Icaronycteris*, was found in early Eocene rocks in a lake in North America and was as perfectly adapted for flapping flight as any modern bat. Its teeth prove that it was an insect eater, belonging to the order of bats now called Microchiroptera. By the end of the Eocene, several groups of modern bats had evolved, including the horseshoe bats. During the day they could not compete

Paramys, an early squirrel-like rodent, was one of the first of many new rodents to evolve in the Eocene. As they spread, they drove primates such as the tarsier-like *Tetonius* from their habitats. Primates just managed to survive in North America and Europe but it was only in Africa and South America that they made important evolutionary changes.

with the many bird species but at night they were able to fly freely and feed undisturbed on insects and fruit. Since they must have been descended from small, nocturnal mammals, darkness would be their natural world and they had probably already evolved their special system of finding their way about by sound.

The plant eaters

As the Eocene epoch progressed, the plant eaters that had evolved during the Palaeocene became very successful. An important one was a condylarth called *Phenacodus* which has been found in both North America and Europe. *Phenacodus* was quite a small animal, about as big as a large dog, with a long head and body and short legs. Unlike the earlier condylarths it had small hooves instead of claws on its toes, possibly to help it to move more quickly.

Two other animal groups descended from the condylarths appeared but also became extinct during the 16 million years of the Eocene epoch. One group, the tillodonts, were about the size of small bears. The other, the taeniodonts were much larger and may have looked rather like giant rats. We do not know enough about these two groups to say why they died out when others survived. For some reason they were unable to compete with the growing number of plant eaters.

Pantodonts continued to live in the lakes and rivers, probably coming out at night to graze. Larger pantodonts now appeared: one, *Coryphodon*, was over 3m long. Its spine was too weak to support it permanently on land – another sign that it must have spent much of its time in the water.

By far the most dramatic looking mammals were the rhinoceros-sized uintatheres. Several different species of these evolved, gradually increasing in size. The bony knobs on their heads became more and more exaggerated until in the giant *Uintatherium* and its 4m long relative *Eobasilus*, they were quite grotesque.

Again and again in animal evolution we find that animals gradually increase in size. Unless there is a positive advantage in being small (to avoid being noticed by predators, for example) most species tend to produce larger and larger forms. This happens because in each generation the larger and stronger animals are more likely to survive for longer and so produce more offspring. Larger animals have fewer enemies and as long as there is enough food to support them, they continue to be successful.

The history of the uintatheres shows this very clearly. The first, tapir-sized animals gradually developed into bigger and bigger forms until by the time they died out

Phenacodus

they were more than 2m high at the shoulder and up to 4m long. Why they finally became extinct is not known.

The odd-toed ungulates

Odd-toed ungulates are animals with an odd number of hoofed toes on each foot. (Ungulate means simply 'hoofed'.) Today they are the horses, tapirs and rhinoceroses.

The very first odd-toed ungulate was found in England in 1840 by a man named Richard Owen. He named the fossil *Hyracotherium* because he believed it was related to modern hyraxes. *Hyracotherium* stood about 20cm high at the shoulder and had a long head and body. It was probably descended from condylarths like

Coryphodon, the largest of the pantodonts, was over 3m long and lived like a hippopotamus in lakes and rivers. The amynodonts (*Metamynodon*) were early members of the rhinoceros family. They too were water loving animals and in time took *Coryphodon's* place. *Phenacodus* was a condylarth, a plant eater. It is the first known mammal with nail-like hooves on its toes instead of claws and may be a forerunner of the first horses. It was about the size of a large dog.

Metamynodon

Coryphodon

Phenacodus but it had longer legs and ran on its fingers, not on the flat of its hands and feet. Its fingers and toes had small hooves and behind them was a fleshy pad which took the animal's weight as it ran. The front feet had four fingers and the back only three. From the size of its skull we know that its brain was probably well developed.

Much later, fossils of the same animals were found in North America and there it was named *Eohippus* or Dawn horse because scientists realized that it was the ancestor of the modern horse. The name *Hyracotherium* is still used because it was proposed first and although it is less appropriate, a change would have been confusing after such a long time.

Next page: The Eocene was dominated by odd-toed ungulates, including the earliest horses (*Hyracotherium* 1), rhinoceroses (*Hyrachyus* 2) and tapirs (*Helaletes* 3). There were also giant plant eaters such as *Uintatherium* 4 , with bony growths on their skulls and dagger-like tusks. Packs of running, dog-like mesonychids (*Dromocyon* 5) hunted the smaller plant eaters.

In North America two other horses evolved from *Hyracotherium*, *Orohippus* and *Epihippus*. They were larger than their ancestor (between 25 and 50cm at the shoulder) but in other ways very similar. The main change was in the teeth which became better adapted for dealing with plant food. Like *Hyracotherium* they were forest animals, feeding on leaves in the dense undergrowth. In Europe a group of tapir-sized animals called the palaeotheres evolved from *Hyracotherium*, but the true horse line died out.

The first tapirs and rhinoceroses evolved around the same time as the first horses, probably also from condylarth plant eaters. They, too, were small, slender animals that lived in the forest and fed on leaves. They looked very unlike their modern descendants and it is only by studying the details of their teeth that we know they were in fact the basic stock from which the later animals evolved.

One branch of the rhinoceros family, the amynodonts, became large, barrel-shaped animals with short legs, like hippopotamuses. Their eyes were high up on their skull and they probably spent much of their time in the water with only their eyes showing above the surface – exactly as a hippopotamus does today.

Two important but now extinct groups of odd-toed ungulates first appeared in the Eocene, the brontotheres in the New World, and the chalicotheres in the Old. The earliest brontotheres evolved in both western North America and eastern Asia but later ones are found only in North America. The earliest were only about the size of modern tapirs but they, like so many other animals, produced larger and larger descendants.

The chalicotheres survived throughout the Age of Mammals and in fact only died out in the last two million years. At first they probably looked much like the other early odd-toed ungulates for at that time the animals were only just beginning to evolve the special features which were to separate them into distinct families. Their most obvious difference was that they had claws instead of hooves. Later they developed horse-like heads and long legs. No-one knows exactly how the chalicotheres lived but as they survived for such a long time, they must have found a successful way of life. Some people think they used their claws to pull down the branches of trees so that they could feed from leaves that other animals could not reach. Others think they may have used them for digging up roots.

The Eocene epoch was by far the best time for the odd-toed ungulates. They were the most successful plant eaters and filled most of the available ecological niches. Many animals developed that are no longer living today

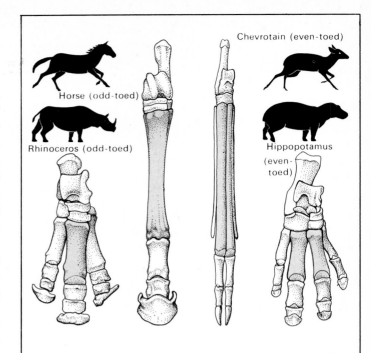

Odd and even-toed ungulates
Both groups of ungulates include heavily and more lightly built animals. Although their leg and feet bones are arranged differently, they share many of the same adaptations. Heavy hippopotamuses (even) and rhinoceroses (odd) have short feet from ankle to toes and take their weight on several toes. Lighter faster animals such as the chevrotain (even) and horse (odd) have longer leg bones and run on one or two toes only. (Illustrations not to scale.)

and although the horses, rhinos and tapirs have lasted, the group was never again so varied.

The even-toed ungulates

The even-toed ungulates include pigs, peccaries, hippopotamuses, camels and llamas, deer and antelopes of all kinds, cattle, goats, sheep and giraffes. Even-toed ungulates have an even number of toes on each foot, often divided into a cloven hoof. Their leg and foot bones are arranged differently from those of odd-toed ungulates and they are more specialized for springing and leaping than for running. Their teeth are also different, and are well adapted for crushing and grinding their food.

Like other plant-eating mammals, the even-toed ungulates probably evolved originally from the condylarths and they first appeared as a separate group at the beginning of the Eocene some 54 million years ago. The first pigs, camels and the ancestors of chevrotains, the first primitive ruminants (cud chewers) come from this time. Most of the early even-toed ungulates probably looked and behaved rather like pigs.

Later they began to adapt to follow different ways of life: one took to the water, another had claws and seems

Patriofelis

Andrewsarchus

Dromocyon

Eocene flesh eaters. *Andrewsarchus* was a mesonychid, a member of the same dog-like group as *Dromocyon*. However, it was too large to be a fast hunter and probably lived more like a modern bear. Only its metre long skull has been found. No-one knows exactly what its body looked like but from the evidence of its teeth and skull scientists can reconstruct the type of size and shape it must have been. Oxyaenids such as *Patriofelis* were stealthy, cat-like hunters but were not the ancestors of modern cats.

to have lived in the same way as chalicotheres but in North America instead of Europe. Before the end of the Eocene all the main groups of this order had evolved.

The flesh eaters

The most important carnivores during the early part of the Eocene were the flesh-eating birds that had been the mammals' main enemy since the extinction of the dinosaurs. By the middle of the Eocene, however, many plant-eating mammals were becoming larger and flesh-eating mammals became the dominant predators.

In the Palaeocene epoch five groups of carnivores had evolved: the dog-like mesonychids, the bear-like arctocyonids, the cat-like oxyaenids, mongoose-like hyaenodonts and the small miacids. All these groups continued to flourish in the Eocene.

The dog-like mesonychids now became specialized for running, moving on their fingers and toes instead of on the flat of their hands and feet like most mammals of the time. They were the 'dogs' of the Eocene and, like true dogs, they probably hunted in packs, easily outrunning the more slowly moving herbivores. One mesonychid, *Andrewsarchus*, grew into a giant of its kind:

its skull alone was about a metre long. Such a large animal was obviously not a runner and it probably fed on both flesh and plants like a modern bear. In North America the arctocyonids lived in the same way.

The cat-like hunters, the oxyaenids, do not seem to have been adapted for running. Instead their strong front legs and powerful shoulders show that they could pounce on their prey, probably taking it by surprise in the forest undergrowth. These, like the dog-like flesh eaters, produced giant versions.

The fourth group, the mongoose-like hyaenodonts, remained small and lightly built, ranging in size from polecats to foxes. With their long, low bodies and long tails they hunted in the dense undergrowth of the forests. In North America there was a sabre-toothed hyaenodont with long, stabbing teeth, very like the real sabre-toothed cats that evolved later from a different ancestor.

The ancestors of both true cats and true dogs were the small tree-living hunters, the miacids. In the Eocene they made few major evolutionary changes, though they were beginning to separate into the two main groups from which dogs and cats were to evolve.

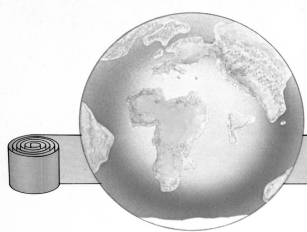

Mammals in the Oligocene Epoch

The Oligocene epoch lasted for fourteen million years (from 38 to 26 million years ago). During this time the geography of the world continued to change. At the end of the Eocene, Europe had been separated from both North America and Asia by sea. During the Oligocene the Earth's axis of rotation tilted slightly and an ice cap formed at the South Pole. Because a large quantity of water was locked up as frozen ice, sea levels fell all over the world and the shallow sea that had separated Europe and Asia disappeared. They now formed a single land mass across which the mammals could migrate easily.

The southern ice cap affected the world's climate and as a result the vegetation also changed. The dense forests that had covered most of the lowlands gave way to more open woodland with clumps of trees and clearings.

For the mammals it was a period of slow but steady change. True rodents became well established: the squirrel *Sciurus* which still lives in North America and Europe had appeared by the start of the Oligocene some 38 million years ago, and there were early species of rats, mice, hamsters, voles and dormice. The first known beaver comes from the early Oligocene of North America.

The spread of tree-living rodents finally drove the primates from all the northern continents. Fortunately some had reached South America and Africa and there they continued to evolve.

Rabbits and hares had already appeared in Asia and North America and now the first pikas, the second group of the lagomorph order, evolved – apparently from the same common ancestor. The first fossils come from Asia, but pikas quite soon spread to Europe and across to North America. Unlike rabbits and hares, they have rounded bodies with no visible tail and short, round ears set close to their heads. These may be an adaptation to a colder climate, for pikas now live in high mountain areas where it is important not to lose heat from the body's surface. At this time there were no hares in Europe. In their place, 30cm long hopping or jumping cainotheres evolved – early relatives of the camel group.

The odd-toed ungulates

In the Eocene the odd-toed ungulates had developed into a very varied group. Several dramatic changes now took place.

In Europe and Asia the early horse *Hyracotherium* and its descendants the palaeotheres died out completely. In North America, however, horses flourished and began to grow larger. The two small Eocene horses *Orohippus* and *Epihippus* died out but their descendant *Mesohippus* survived. *Mesohippus* looked much more like a modern horse than its ancestors, though it was still on a miniature scale, about the size of a greyhound. A slightly larger horse, *Miohippus*, followed it.

By far the most sensational looking of the odd-toed

The first rabbits and hares had appeared in Asia by 38 million years ago and soon reached North America. Pikas also first appeared in Asia and spread to Europe and North America. Hares did not reach Europe at this time. Their ecological niche was occupied by small jumping cainotheres, close relatives of the camel.

Brontotherium

Rabbit

Pika

Cainotherium

42

ungulates were the brontotheres which evolved mainly in North America. By quite early in the Oligocene they had grown from tapir-sized animals to giants such as *Brontotherium* which stood nearly 2m high at the shoulder. As they increased in size they developed large, bony knobs on their snouts. *Brontotherium's* bony knobs were so large that they rose higher than the back of its head.

The brontotheres were in many ways like the large plant-eating uintatheres and gradually took over their place in the environment, driving them into extinction. By the end of the Oligocene the brontotheres were extinct in their turn

Several different types of rhinoceros were alive at this time, some of them becoming very large indeed. A group of giant hornless rhinoceroses in Asia included the largest land mammals that have ever existed. *Baluchitherium* was 8m long, stood 5m at the shoulder and weighed about 16 tonnes, eight times as much as the largest living Indian rhinoceros. They had long, pillar-like legs to support their heavy bodies and long necks. Like modern giraffes they could reach up to feed on the higher branches of trees and so avoid competition for food among the lower branches. This type of rhinoceros

Next page: The Oligocene was a time of change and even-toed ungulates began to replace odd-toed as the most important plant eaters. The first modern flesh eaters, the cat *Dinictis* (1) and dog *Pseudocynodictis* (2), appeared. Of the primitive flesh eaters only the scavenging hyaenodonts (*Hyaenodon* 3) continued to flourish. Among the even-toed ungulates the first cud-chewers or ruminants (*Protoceras* 4) evolved. Animals like *Protoceras* were the forerunners of deer and giraffes as well as of cattle, sheep, antelopes and their relatives. Enteledonts or giant pigs (*Archaeotherium* 5) were not related to true pigs and were about the size of modern cattle.

During the Eocene and Oligocene a group of large odd-toed ungulates, the brontotheres, evolved in North America. About the size of modern rhinoceroses, they had a forked bony 'horn' on the front of their snout. In Asia true rhinoceroses evolved a giant hornless form, *Baluchitherium*. The largest land mammal ever known, it stood 5m at the shoulder and was 8m long.

Baluchitherium

Brontotherium

has only been found in eastern Asia and does not seem to have reached either Europe or North America.

In North America an entirely different type of rhinoceros, *Hyracodon* evolved. *Hyracodon* was about the same size and shape as the early horses of its time and must have competed directly with them for food. The hippo-like rhinoceroses that had appeared in the Eocene continued for a time to live in the rivers and lakes of North America, Asia and Europe and in North America and Europe the direct forerunner of modern rhinos, the hornless *Trigonias* evolved. By the middle of the Oligocene, about 42 million years ago the two-horned Sumatran rhinoceros which still survives today was alive.

The even-toed ungulates

In spite of the large brontotheres and rhinoceroses and the gradually developing horses, most odd-toed ungulates were not doing very well in the Oligocene. It was another group of hoofed plant eaters, the even-toed ungulates, that were beginning to expand.

The even-toed ungulates are divided into two main groups, the pigs and the ruminants or cud-chewers. Pigs are described as root eaters but in fact they will eat virtually anything they can find and act as general scavengers. Forerunners of the pigs had evolved in the Eocene but little is known about them until the Oligocene when they had already developed many of the characteristics of modern pigs. From then on they gradually increased in size and adapted to many different forest environments. Today, even without the many domestic varieties, there are pigs of various species living in Europe, Asia and Africa. Until they were imported by man there were no pigs in North America where a group of similar animals, the peccaries, took their place.

In both Europe and North America there was a group of animals called the entelodonts. These are often called giant pigs though in fact they are not related to true pigs at all. About the size of modern cattle, they had large, heavy heads and strong muscular shoulders which gave them a rather humpbacked look. By contrast, their legs were long and slender and they could probably run quite fast. Although their skulls were large, their brains were apparently small and this is probably why they were eventually driven out by the more intelligent pigs.

Another early group of even-toed ungulates, called anthracotheres, evolved in Asia and spread into Europe. They belonged to the pig group but were specialized for living in and beside lakes and rivers. In time they reached North America and there they drove out the water-living rhinoceroses. Anthracotheres died out in

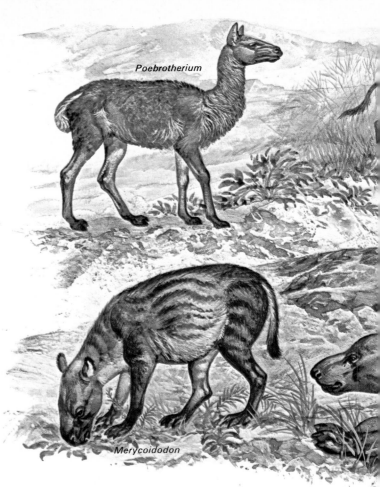

Thirty million years ago oreodonts, a group of even-toed ungulates that are now extinct, were among the most important plant eaters. Some, such as the pig-sized *Promerycochoerus,* were water dwellers; others, such as *Merycoidodon* (the size of a peccary) lived on land. Another early even-toed ungulate, *Agriochoerus,* was built more like a flesh eater with powerful bones, a long tail and claws. Its teeth, however, prove that it was a plant eater. Like the oreodonts and the sheep-sized camel *Poebrotherium,* it may have been a primitive cud-chewer. Most odd-toed ungulates were now beginning to decline but rhinoceroses were still developing new forms. *Hyracodon,* a running rhino, was similar to the horses of its time.

most places during the Miocene period but scientists think that modern hippopotamuses may be descended from some of their later forms.

Another group of even-toed ungulates seem to have been the most widespread of all mammals in the Oligocene. These were early ruminants or cud chewers called oreodonts. More fossils of oreodonts have been found from this time in North America than of any other mammals. This may be partly because they lived near water, where conditions were ideal for fossils to form but there are so many bones preserved that they must have been there in very large numbers.

The other early ruminants that evolved in the Oligocene, the ancestors of camels, giraffes and deer, were still very different from their modern descendants but animals looking almost exactly like today's chevrotains

Hyracodon

Agriochoerus

Promerycochoerus

hyaenodonts died out, too, but they left descendants which lived as flesh-eating scavengers and looked generally like modern hyaenas. They probably sometimes hunted down their prey in packs but relied mainly on the remains of other animal kills. These hyaenodonts lasted for thirty million years, spreading across North America, Europe and Asia and even reaching the isolated African continent. Here they were for a long time the most important flesh eaters and must have been active hunters.

The animals whose descendants were to become the main flesh eaters of the Oligocene were the miacids, the long-bodied tree-living mammals that had evolved at the same time as the other early flesh eaters but had until this time seemed less important.

Modern flesh eaters are divided into two groups, the cats and the dogs. We know exactly what the earliest true cats and dogs looked like because some of the same groups of animals are still alive today. The first cats included civets and mongooses; the first dogs were small polecats, weasels and similar long-bodied hunters.

Around the middle of the Oligocene, about 30 million years ago, two basic lines of true cats appeared – the stabbing and the biting cats. The biting cats were lightly built, adapted for running and pouncing on their prey, relying on speed and surprise. They had short, strong canine teeth and powerful jaws which bit through the neck veins easily. The stabbing cats were more heavily built and probably hunted the larger and slower plant eaters. They evolved long, dagger-like canines with which they stabbed and slashed their prey to death.

The dog group of carnivores includes not only dogs, foxes and wolves but also bears, pandas, badgers and the smaller weasel-like group. *Mesocyon*, the ancestor of true dogs and wolves lived in North America during the Oligocene and there were also other important groups. *Amphicyon*, a huge dog the size of a grizzly bear, lived all over the northern continents and survived until about seven million years ago. The ancestor of raccoons, also members of the dog family, evolved from slender animals that looked more like primitive cats than dogs; and otters appeared in Europe.

At the end of the Oligocene, then, the direct ancestors of all the modern groups of flesh eaters were well established. The plant eaters they preyed on were still at an earlier stage of evolution: a few were becoming modern but many primitive groups were still living successfully. If people had been alive 26 million years ago, they would have found it hard to believe that the even-toed ungulates were soon to become established all over the northern lands as the dominant plant eaters.

were already living in North America over thirty million years ago.

Most ruminants today have horns, some of them forming spectacular outgrowths from their heads. Their Oligocene ancestors were hornless but one relative of the small chevrotains, *Protoceras*, shows the first signs: males had two flat bony growths on the snout and club shaped knobs behind the eyes.

The flesh eaters

Plant eaters during the Oligocene were changing only very gradually into more modern-looking animals. With the flesh eaters the changes were, in the evolutionary sense, much more abrupt. During twelve million years the cat-like oxyaenids, dog-like mesonychids and bear-like arctocyonids all vanished. The early mongoose-like

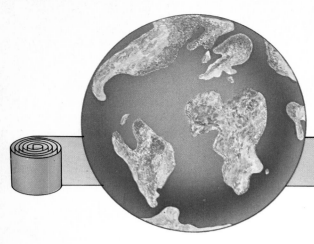

The Southern Continents

While mammals were slowly changing into their more familiar modern forms in the northern continents, animal life in the southern land masses was quite cut off from the mainstream of evolution. Australia became separated very early on, before placental mammals had evolved at all, but by the time South America and Africa were cut off, several groups of these more adaptable mammals were alive. It was from this basic stock that all the mammals of Africa and South America evolved.

Although the land masses were isolated, their climate and vegetation remained for a long time much the same as it was on the other continents. It is not surprising, then, that African and South American mammals often made many of the same adaptations as the mammals in the north. Sometimes they looked physically different, even though they were living in the same ecological niches as their northern equivalents. Sometimes quite unrelated groups even looked alike because they had both adapted in the same way to the same special environment. American anteaters and African aardvarks are good examples among living animals. Both live in much the same way, digging with powerful claws at night for ants, worms and other insects. Both have long, probing noses and a long, sticky tongue which licks up food. This type of similarity in unrelated animals is said to be the result of convergent evolution.

MAMMALS IN AFRICA

Condylarths, primates, rodents and small insect eaters had all evolved before Africa was cut off from the other continents, between 65 and 55 million years ago. It is very likely that these widespread and successful animals were living in Africa at the time, as well as in other parts of the world but so far not a single mammal tooth and not even the fragment of a bone from the Palaeocene has been discovered anywhere on the continent.

The first whales

The earliest African mammal fossils are about 46 million years old and are all of water-living mammals, including the first known whales.

The early whales had short streamlined bodies rather like modern dolphins, but one, *Basilosaurus*, was like a giant sea-serpent, over 20m long. The animals whose fossils have been found were certainly not the first of their line. From their teeth and parts of their skull scientists believe they were descended from the mesonychids, the dog-like hunters that had evolved in the Palaeocene while Africa was still linked to the other continents. There is no evidence yet for the long series of evolutionary changes that must have taken place to produce fish-eating mammals, fully adapted for life in the water and no longer able to live on land at all.

Today's whales are divided into two groups, toothed whales which have a large number of teeth to trap fish inside their mouths; and whalebone or baleen whales,

Palaeoparadoxia   *Desmostylus*

Around 50 million years ago mammals returned to the seas. The early whales included the great sea serpent *Basilosaurus*. Dugongs and manatees evolved around 45 million years ago and along the coasts of the North Pacific the desmostylans appeared. Some, such as *Desmostylus*, looked similar to modern walruses but one, *Palaeoparadoxia*, was more like a sea hippopotamus.

Manatee

which have no teeth but curtains of fine mineralized hair or baleen in their mouth. This filters tiny shrimps from the surface waters of the ocean. Both groups seem to be descended from the early African whales and by the end of the Eocene, 38 million years ago, had spread to the waters around North America and Europe. The first baleen whales come from 45 million year old rocks that are now in India and have also been found in New Zealand. The first toothed whales come from rocks of about 40 million years ago in North America. These rocks are now part of the continents but when the whales died they formed part of the ocean floor.

At the time the whales were evolving, much of Africa was covered with swampy forests, an ideal environment for water animals. Once they had become fully adapted to life in the water, the whales were able to spread widely throughout the oceans, unaffected by the separation of the land masses. Today they are a very varied group. They include the blue whale, the largest animal ever to have lived and the dolphin, whose brain is almost as well developed as our own.

The sirenians – the dugongs and manatees – are the only other living group of fully aquatic mammals. They are probably descended from plant-eating condylarths and their ancestors have been found in rocks about 45 million years old in Egypt, Libya, Somalia, Jordan, India and Jamaica. In some cases only fragments of their ribs have been found but they are easy to identify as their bones are heavy and solid, not spongy like those of other animals. The first known sirenians, like the first whales, were already very different from their land-living ancestors. They were well adapted for life in the water with flippers instead of limbs and a streamlined shape for effective swimming.

Another group of water-living animals, the desmostylans, lived like sea hippopotamuses along the coasts of the north Pacific ocean. Their fossils have not been found in Africa but they are included here because they may belong to the same group as both sirenians and early elephants – both of which certainly began in Africa.

Mammals on the land

The earliest mammal fossils in Africa are all of water-living animals. This does not mean that there were no land mammals at the time, simply that all the rocks of the right age that have been searched for fossils were formed from marine deposits.

In the last forty million years Africa's climate and

Basilosaurus

vegetation have changed dramatically. When the early mammals were alive it must have been a semi-tropical forested country with swampy pools surrounded by thick stands of trees. There were no open plains and no desert areas. Many of the mammals must have lived in the north, where the great Sahara desert now lies and in the central regions where thick forest makes exploration impossible. In places where sedimentary rocks are exposed, the action of wind and rain has often worn all evidence away to dust.

It is easy to forget the vast ages of time that have passed since the earliest mammals were living: the end of the Eocene, when the first fossils of land mammals are found in Africa, was nearly 40 million years ago. Remembering this, it is more surprising that any fossils have been found at all than that there are so few.

It is not until the Oligocene (38 to 26 million years ago) that there are enough fossils of land mammals to give a clear idea of what the animals looked like and how they must have lived.

The African animals that were closest to their condylarth ancestors were members of the hyrax family. Today the only members of the family are the small conies and dassies but in the Oligocene there were several different types. These ranged from *Saghatherium*, the size of a large cony, to the sheep-sized *Megalohyrax* (which looked rather like an early horse) and a rhinoceros-sized giant named *Titanohyrax*.

Another large plant-eating mammal found in Africa was *Arsinotherium*. *Arsinotherium* is a rather mysterious animal. It looked rather like a modern rhinoceros but was not related to the same group. No other animals of its type have been found and although its teeth link it in some ways to the hyraxes, other characteristics set it apart. We know nothing about its ancestors, except that they were most probably a group of condylarths, and it has left no known descendants.

The elephants, which appeared for the first time about 40 million years ago, are much better known, though again their early history is guesswork. *Moeritherium*, the first real elephant, has been found in Egypt, Senegal, and Mali. It was about 60cm at the shoulder and had a long body with short legs, rather like a pigmy hippopotamus. Like a hippopotamus it probably lived in swampy areas. It had a short, mobile snout instead of a trunk. Fragments of the skull of another elephant-like animal, *Barytherium*, have also been found. *Barytherium* seems to have been much larger than *Moeritherium*, about the same size as an elephant today. However, very little is known about it and it is from *Moeritherium* that modern elephants evolved. By the Oligocene larger

Arsinotherium was a large plant eater, about the same size as a rhinoceros. It has been found only in Africa and although we know from fossil skeletons what it looked like, we know very little about its ancestors and it has no known descendants. The main plant eaters in Africa were the hyraxes. *Saghatherium* was about the size of a modern hyrax but *Megalohyrax* was as big as a sheep.

elephants, some as much as 2m tall, were alive.

Sirenians, desmostylans, hyraxes, elephants and *Arsinotherium* are sometimes grouped together as subungulates. (Subungulates are 'not quite' ungulates: they have thick nails instead of hooves on their toes.) Except for the desmostylans, their early forms are found only in Africa. Three other mammal orders, the insect eaters, the rodents and the primates also developed differently in Africa from the rest of the world.

The first African insect eaters evolved from animals that had been living in the continent for millions of years, at least since the beginning of the Tertiary period, 64 million years ago. One primitive animal named *Ptolemaia* may have been living there in the Cretaceous when the dinosaurs were still alive. A group descended from these insect eaters, the elephant shrews, are still found only in Africa.

Several rodents appeared in the Oligocene in Africa. They included cane rats, primitive but still living Old World porcupines. Cane rats are now serious pests to food crops in tropical Africa but are also an important – and delicious – food themselves.

The primates

The third order of mammals to develop independently in Africa was the primates. Like the rodents and insect eaters they had begun to evolve while Africa was still linked to the north. There, the primates were driven out

Arsinotherium

Megalohyrax

Saghatherium

by rodents but in Africa and South America they continued to flourish.

Two separate evolutionary lines developed at the same time in Africa, both specializing in different ways for life in the trees. The first were the Old World monkeys or catarrhines, which include macaques, baboons, langurs and others. The second were the hominoids – great apes, gibbons and man.

Fossils of the early ancestors of these main primate lines come from the Oligocene rocks of Africa: the ancestors of Old World monkeys, of gibbons and of the ape and man group. Nobody knows why these important primates evolved only in Africa but there can be no doubt that it was here that the line leading eventually to our own species began.

The immigrants

Among the few fossils of land mammals from the Eocene are some of the northern hyaenodont flesh eaters and pig-like anthracotheres. Somehow these managed to cross from Europe or Asia where both groups later became extinct. In Africa the hyaenodonts became the most important flesh eaters while the anthracotheres were ancestors of modern hippopotamuses. In spite of these important immigrants, African mammals continued to develop in their own unique way. It was not until the Miocene that the influence of northern mammals became really significant.

MAMMALS IN SOUTH AMERICA

At the beginning of the Age of Mammals South America, like Africa, became isolated from the other continents. Even the narrow land link with North America was broken. As in Africa, several primitive mammal groups were already living there and also as in Africa, new animals evolved from these basic stocks to fill the available ecological niches.

In the north, different groups of animals constantly replaced one another as they competed for food and living space, often migrating to find new areas and adapting to use new types of food. After thirty million years, African animals, too, were once more in competition with other groups when Europe and Asia were linked to the southern continent again. But South America continued to be completely isolated until the Pleistocene, two million years ago. For almost sixty million years the animals were able to develop undisturbed. If you look back at the enormous changes that took place in even the first ten million years of the Age of Mammals you will be able to imagine how different the animals in South America became.

Back in the Palaeocene, 64 million years ago, very few mammals were alive in South America. There were early marsupials, two groups of placental hoofed plant eaters and placental armadillos. The placental insect eaters that were so widespread in Asia and North America had not reached South America and there were no placental flesh eaters. There were no even- or

Next page: During the Miocene the developing Antarctic ice cap brought seasonal snows to southern South America. Among the plant-eaters were litopterns, including the camel-like *Theosodon* (1) and the horse-like *Thoatherium* (2), armadillos (3) and ground sloths (4). The flesh eaters were all marsupials. Borhyaenas (5) were the major predators but more lightly built animals such as *Cladosictis* (6) hunted typotheres (*Pachyrukhos* 7) and rodents (*Eocardia* 8).

odd-toed ungulates and none of the large pantodonts and uintatheres that evolved in the north.

The marsupials

With no placental flesh eaters to compete against, the marsupials adapted to fill all the carnivore niches – insect eaters, scavengers and active hunters of many different sizes.

Insect-eating opossum rats which still survive today in Ecuador, Columbia and Peru, first evolved in the Palaeocene. Sixty million years ago they were just one of a large number of different types of marsupial insect eaters including a group very like placental rodents and another like the multituberculates.

The first opossums came from North America in the Cretaceous but by the Palaeocene had spread to both Europe and South America. They died out in Europe but have continued to flourish in North and South America right up to the present day. Opossums are the least specialized and most primitive of all marsupials but like shrews and hedgehogs they have survived for more than 64 million years, competing successfully with apparently better adapted animals. They are voracious feeders and will eat almost anything that comes their way – insects, small mammals and birds. Even now they are extending their range in North America, where they are becoming city scavengers, feeding on garbage left in the streets and around houses.

Opossums were active flesh eaters but they were not the main threat to plant eaters in South America. These were animals called borhyaenids, marsupials descended from opossum stock. The animal that gives the group its name is *Borhyaena*. About the size of a wolf, it was heavily built with a short snout that made it look rather like a cat. Its powerful limbs show that it could pounce in attack but it may also have been a scavenger, feeding on animals that had already died or been killed. Borhyaenids developed in several different ways from small, long-bodied animals like polecats and civets to giant bear-like creatures such as *Arminiheringia*. Much later, in the Pliocene (7 to 2 million years ago) a marsupial sabre-toothed cat evolved named *Thylacosmilus*. Its tail was short and its limbs very powerful. At the front of its mouth its two upper canine teeth curved down like long daggers, sliding into a skin sheath at either side of its jaw when its mouth was closed. *Thylacosmilus* must have lived and hunted just like the placental sabre-toothed tigers of the northern continents. However, it evolved quite separately and, like all the other South American flesh eaters, was a pouched mammal with a marsupial's method of reproduction.

The plant eaters

Condylarths were living in South America before its isolation and as in the other continents, more specialized plant-eating mammals evolved from them. In South America these were the litopterns.

Two main lines of litopterns evolved in the Palaeocene. The first were similar to early horses. They had long, slender limbs and were built for running. There was a three-toed litoptern about the same size as the true horse *Mesohippus* and later, in the Miocene, a one-toed litoptern with leg bones very like those of modern horses. None of this horse-like group grew very large.

The second line of litopterns also had long legs but with their longer necks they looked like modern llamas.

The main plant-eating mammals of South America were not the litopterns but another group of hoofed animals called the notoungulates or southern ungulates. Although their ancestor lived in the north, from the time South America became separated they have been found only in the south. There they are the equivalents of the odd- and even-toed ungulates.

The best known of the early notoungulates is an animal named *Thomashuxleya*. It was 1.5m long with a long body and a short tail. Its head was heavy and was set on a short, thick neck. In many ways it looked like the North American oreodonts – the cud-chewing plant eaters whose fossils are so abundant.

From animals like this a large and important group, the toxodonts, evolved. Toxodonts were not related to the northern rhinoceroses but because they adapted to very similar environments, they often evolved in the same ways. *Adinotherium*, an early toxodont about the size of a sheep but with shorter legs, was the equivalent of the small running rhinoceroses while *Nesodon* was like the more heavily built forms.

One group of toxodonts in the early Oligocene, the homalodotheres, evolved in a similar way to the chalicotheres of the north: they had long front legs and sloping shoulders and backs. Their hands had four fingers, ending in sharp claws. Like the chalicotheres these may have been for digging in the ground but could also have been used to pull down foliage from trees.

Two smaller and more lightly built notoungulates evolved in the Palaeocene, at the same time as the more heavily built toxodonts. One, the typotheres, were similar to ground-living rodents and had the same kind of chisel teeth. The smallest notoungulates of all were the hegetotheres, rabbit-sized creatures with stumpy tails and long back legs which enabled them to run and jump like hares.

In the early Eocene a strangely proportioned animal

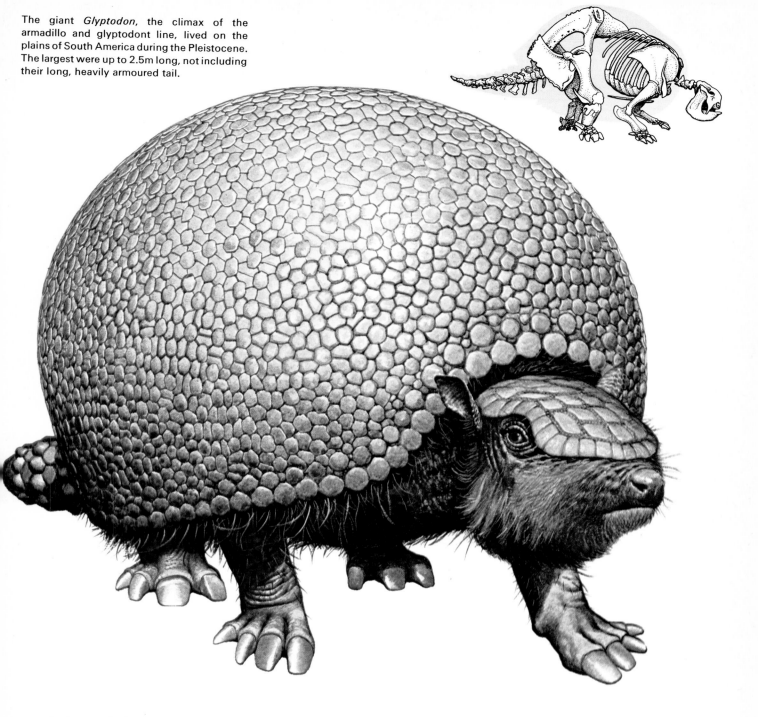

The giant *Glyptodon*, the climax of the armadillo and glyptodont line, lived on the plains of South America during the Pleistocene. The largest were up to 2.5m long, not including their long, heavily armoured tail.

named *Astrapotherium* evolved. Like so many animals throughout the Age of Mammals, it was adapted to a hippopotamus-like life, its heavy body partly supported by the water.

Another South American mammal from the middle Eocene and early Oligocene was so like an elephant that for a long time scientists believed it must be a direct relative. *Pyrotherium* had long chisel-like tusks, a long trunk and teeth very like those of early elephants. In spite of this, scientists now believe that it evolved quite independently – a perfect example of convergent evolution. By the Oligocene, *Pyrotherium* was already as big as a modern elephant but while the true elephants were beginning to change into their familiar form in Africa,

the South American equivalent died out, leaving no descendants.

The edentates

The evolution of the carnivores and the early placental plant eaters was basically similar in both North and South America – though different families of animals were involved. Most of the main ecological niches in the south were filled by mammals that, at least outwardly, looked like those in the north. But South America also produced a group of mammals quite unlike any in the rest of the world – the edentates.

The word edentate means toothless, though in fact most of them do have simple peg-like teeth at the back of their jaws. Their teeth are different from those of all other mammals and there are also differences in the structure of their backbone. They have extra joints between the vertebrae in the lower part of the back, making it very strong.

There are three major divisions of edentates: the armadillos, the sloths and the giant anteaters.

The first known edentate comes from North, not South America. It was a primitive armadillo named *Utaetus* and probably evolved from early insect eaters. Armadillos are the only mammals ever to have true bony armour – the bulky rhinoceros's defensive shields are formed from tough skin. Bone armour is common enough in turtles, crocodiles, even dinosaurs but in all these animals it was something that remained from an earlier stage of development. For mammals, descended from furry, soft-skinned animals, it was a completely new invention.

Utaetus's armour was made up of hinged plates of bone. If it was threatened, it could roll itself up into a ball and would have looked very unappetizing.

By about 40 million years ago, the armadillos had produced a side branch, the glyptodonts. These armadillos had armour fused into a solid, inflexible bony shell with an extra 'hat' on the top of their head. Although a glyptodont could no longer roll itself into a ball, when it bent its head down into its shell, it was completely protected. Its tail had bony rings instead of a solid covering and could easily be moved from side to side. The last of the glyptodonts were giants named *Glyptodon* and *Doedicurus*. They grew up to 2.5m long with tails of almost the same length again. These long tails must have been effective weapons: armed with groups of short spikes, they could be swung from side to side like the mace of a mediaeval knight.

Today's armadillos are descended from the more primitive group, not from the glyptodonts, which died out less than a million years ago. They are good burrowers and roots are part of their diet as well as fruit, insects, worms and small reptiles. Their legs have become longer and they can run fast. They can also defend themselves effectively with their sharp claws. Like the other edentates they have very small brains and it is probably only because they have become so highly specialized for their particular way of life that they manage to compete with more intelligent animals.

The second group of edentates, the sloths, first appeared in the middle of the Eocene, about 40 million years ago. They were ground sloths with well developed claws on their hands and feet. Perhaps to protect these, they walked on their outer knuckles and on the outer

edges of their feet. Sloths may have been descended from an unknown group of armoured animals for some still had lumps of bone under their skin, the remains of a defensive shell. About two million years ago, in the Pleistocene, ground sloths, like glyptodonts, became giants. The last of the line, *Megatherium*, was 6m long and, standing on its back legs, could have reached high into the branches for its food. *Megatherium* may have survived until only a few thousand years ago for mummified remains with some skin and hair still preserved have been found.

Curiously, there are no fossils at all of early tree sloths, the slow-moving leaf eaters which are the only living members of the sloth family.

The last group of edentates to appear were the anteaters whose first fossils come from the Miocene (26-7 million years ago). Anteaters probably evolved from the same ancestors as tree sloths but have become extremely specialized for an unusual way of life. They are true edentates with no teeth at all and a long, worm-like sticky tongue for catching ants – their staple food. Their snout is very long indeed, with a tiny mouth opening at the end. They have long, powerful claws which they

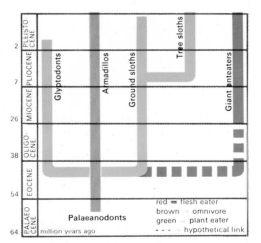

The edentates

The ground sloths which lived throughout the Tertiary period in South America ended with the giant *Megatherium*, up to 6m tall. Living tree sloths are a late sidebranch from the same evolutionary line and none have so far been found as fossils.

use mainly for tearing open ants' nests, though they are also useful for defence. Anteaters today seem to be at their giant stage: their Miocene ancestor *Promyrmephagus* was over a metre long but today's giant anteater is 2m from head to tail.

The immigrants

Although South America was cut off from North America from the beginning of the Palaeocene until about two million years ago, a few mammals did succeed in crossing from north to south, possibly on floating logs and other plant debris. A primitive rodent (perhaps the early squirrel *Paramys*) arrived in South America during the Oligocene and from it the special South American rodents evolved.

These included several rat-like animals, a spiny tree-living ancestor of New World porcupines, fluffy-tailed chinchillas and guinea pigs. A side branch of the small guinea pigs seems to have lived like the first odd-toed ungulates in the north. *Eocardia*, with its compact body and slender limbs, marks the beginning of a line that leads to today's pampas-living capybaras.

At the end of the Oligocene, about 26 million years ago, a second tree-living group of mammals appeared in South America – the primates. Somehow – we do not know exactly how or when – some of the rodent-like primates from the north must have migrated south, perhaps driven from their territory by the true rodents. These were to be the ancestors of the New World monkeys, the platyrrhines.

Unlike the Old World monkeys, the platyrrhines have flat noses with nostrils set wide apart. Today there are two groups, the marmosets, which have thick, bushy tails and climb by digging their claws into the bark; and the more advanced cebids which include howlers, capuchins, spider and woolly monkeys.

Evidence for their evolution is very hard to find and apart from a few fossils from the Miocene we can only guess that they must have evolved in a similar way to the monkeys in Africa. They have not, however, developed so far. They are still tree dwellers and no apes have evolved. Perhaps food was always so plentiful in the trees that there was no incentive to forage on the ground and to make the adaptations that led to man.

During the Miocene yet another group of tree-living mammals reached South America – the raccoons – and in the Pliocene (7 to 2 million years ago) forest-dwelling peccaries and a member of the camel family, the llama, arrived. In spite of these immigrants, South American mammals remained quite distinct from the mammals of the north for millions of years.

The Climax: the Miocene and Pliocene Epochs

The tremendous evolutionary explosion of different types of mammals that had begun in the Oligocene reached its peak in the Miocene (26 to 7 million years ago) and continued right up until the end of the Tertiary.

This remarkable increase in the numbers and variety of mammals, which occurred even in isolated South America, was directly related to changes in the world's climate. The ice cap around the South Pole grew larger, speeding up the fall in temperature. Whereas the average temperature in Europe during the Oligocene had been about 18°C, during the Miocene it was between 14° and 16°C. Today it is only 9°C.

The cooler and drier climate had a profound effect on plant life and grasses spread where forests and woodlands had grown before. In North and South America, Europe and Asia vast regions became grasslands – the prairies, steppes and pampas of today. In Africa savannah grasslands also covered large areas of land that had once been heavily forested. The grasslands were an ideal environment in which mammals could develop, for they can provide food for a far greater number of animals than forest can.

Grasslands not only support greater numbers of mammals, they also provide food for a greater variety of types. Mammals of different sizes feed off different levels of vegetation so that the plants are fully used with as little competition as possible. On the grasslands of Africa there is still a great variety of plant-eating mammals, including browsers feeding off trees and shrubs as well as the more numerous grazers. Even among the grass eaters there is specialization. The zebra, for example, feeds off the coarse heads of the grasses, the wildebeests and gnus off the leaves and the gazelles off the seeds and shoots at ground level. Unfortunately there is no way of knowing whether extinct mammals were specialized in the same way but because of the great variety that evolved in the Miocene, it is very likely that they were.

Life on the open plains brings new problems to the animals. Tough grass is more difficult to digest than leaf protein and teeth and digestive systems must be adapted to cope with it. Both predators and prey are also more conspicuous. Most of the successful herbivores became fast runners while the hunters adapted in various ways. Some relied on speed, others on stealth and camouflage or on co-operative hunting.

The Miocene was also a time of major geographical changes which affected the mammals almost as much as did the altered climate and vegetation. By far the most important event was the northward drift of the continent of Africa. During the Miocene it collided with Europe and Asia and the ancient Ocean of Tethys which had separated Africa and India from Eurasia was cut off from the Atlantic by two strips of land. Africa was no longer isolated from the northern continents and mammals from the north rapidly invaded the south. At the same time several of the mammal groups that had until now evolved quite separately were able to spread northwards.

As Africa and India collided with Europe and Asia the edges of the continents buckled up and folded into huge ranges of mountains – now the Alps and the Himalayas. All that remained of the Ocean of Tethys was an extensive salt water lake and during the next million years this dried out completely. It was not until about 5 million years ago that the Straits of Gibraltar moved apart and over what must have been the widest (16km) and highest (450m) waterfall the world has ever seen poured the waters of the Atlantic Ocean, drowning the land and forming the Mediterranean Sea. From this

During the Miocene the continents of Africa and India collided with Eurasia and the great mountain ranges, the Alps and the Himalayas were formed. Because of changes in the world climate, grasslands spread over wide areas forming the prairies and pampas of the New World, the steppes of Eurasia and the savannahs of Africa. Once the land link was established, animals were able to move freely between Asia and Africa: Africa's long isolation was ended. Many grassland-living animals rapidly spread into Europe and Asia and elephants moved north, even reaching North America. Plant-eating ungulates and their predators the cats and dogs moved southward to Africa. The pattern of animal life established there still remains today.

time, migrating animals had to move from Africa to Europe via the Middle East and to Pakistan and India via Arabia. As the Mediterranean flooded, mammals were forced onto higher ground, surviving on small islands scattered throughout the sea.

On the other side of the Atlantic, South America was joined to North America during the Pliocene, but the land link was thickly forested and few mammals managed to cross it at this time. Further north there was still a land bridge between North America and Asia across the Bering Straits but here the climate was very cold and only mammals already adapted to life in the lower temperatures of the northern lands were able to cross. Tropical and semi-tropical mammals were effectively barred.

During the 24 million years of the Miocene and Pliocene epochs, mammals were more varied than they have ever been before or since. All the modern groups of hoofed mammals evolved, many including far more different species than are alive today. Pigs and hippopotamuses, rhinoceroses, deer and giraffes, antelopes, cattle, sheep and goats, oryxes, gnus and camels all evolved into their modern forms during this time. Horses continued to progress towards today's species and chalicotheres continued to flourish. Several types of elephants spread from Africa to Europe, Asia and across the Bering land bridge to North America. The primates left the forests and took the first step along the line lead-

ing to man. And rats and mice multiplied spectacularly in numbers and species.

Among the flesh eaters were biting and sabre-tooth cats and the first of the modern cat line. True hyaenas evolved and, in the Pliocene, the first true dogs and bears. Like the plant eaters, there were many species of both cats and dogs that have since become extinct: they, like the other mammals, were never again to be so varied.

The world's climate continued to grow colder during the Miocene and Pliocene epochs. At the end of the Pliocene, about 2 million years ago, it began to deteriorate more rapidly as the great Ice Ages approached. As you will see, this was to have an enormous effect on most of the mammal species.

The odd-toed ungulates

The odd-toed ungulates were at their most successful during the early part of the Age of Mammals, particularly in the Eocene. By the beginning of the Miocene, 26 million years ago, only four groups remained: horses, rhinoceroses, chalicotheres and tapirs. The brontotheres, including the giant *Brontotherium*, had all died out.

All the odd-toed ungulates were plant eaters and most lived in densely wooded areas where they fed off the leaves of trees and bushes. Only the rhinoceroses (and the extinct brontotheres) had grown to giant size and adapted to more specialized niches.

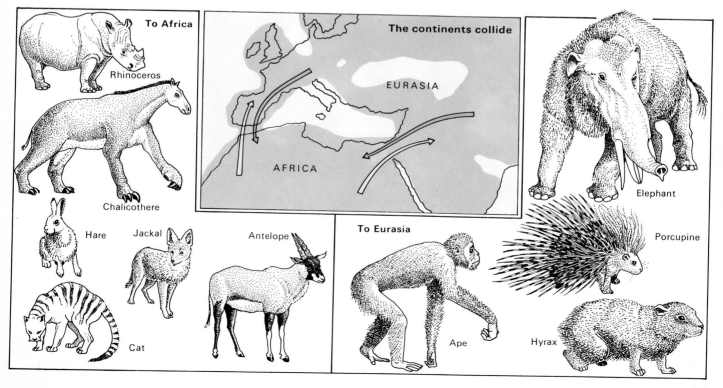

To Africa
Rhinoceros
Chalicothere
Hare
Jackal
Cat
Antelope

The continents collide
EURASIA
AFRICA

To Eurasia
Ape
Hyrax

Elephant
Porcupine

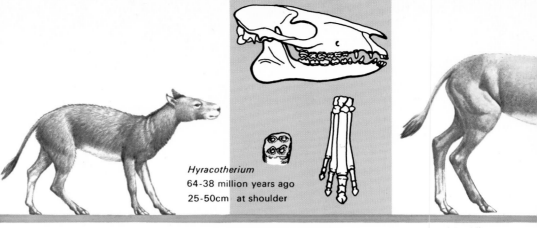

The most obvious evolutionary change in the horse was its increase in size. The first were no bigger than large dogs. By 2 million years ago they were the size of modern ponies while today they range from giant Shires to miniature Shetlands.

Hyracotherium
64-38 million years ago
25-50cm at shoulder

THE EVOLUTION OF THE HORSE

The development of the horse from the small Eocene *Hyracotherium* (the 'dawn horse') to *Equus*, today's genus, is probably the best known of all evolution stories. Fossils of animals at different stages of development have been found and the links between animals separated by millions of years can be clearly traced. So, too, can the environmental changes that influenced the success or failure of each development.

The very first horse, *Hyracotherium*, lived in both Europe and North America in the Eocene but the main line of evolution took place in North America. From time to time, when sea levels were low enough to expose land links, early horses migrated to Asia and Europe.

Hyracotherium was about 20cm at the shoulder, a small, lightly built forest animal with a rounded back. It had four toes on its front feet and three on the back, all with small hoofs. The third toes were stronger than the others and bore most of its weight. It lived in swampy forests, feeding on soft leaves. Its cheek teeth, with four rounded knobs or cusps, were perfectly adequate for dealing with them. The two other horses that evolved in the Eocene, *Orohippus* and *Epihippus*, had teeth that could crush tougher leaves and lived in a drier, but still wooded area.

Horses gradually increased in size throughout the Oligocene when *Mesohippus* and then *Miohippus* evolved. Both were still small compared with modern horses and still walked on three toes on each foot. Living in more open country than their ancestors, they were beginning to run and trot like modern horses.

Several three-toed browsing horses evolved as side branches to the main evolutionary tree. One, *Parahippus*, may have linked *Miohippus* with its later descendants.

For over 12 million years, since the earliest horse *Hyracotherium* and its descendants the palaeotheres had died out in Europe, horses had been living only in North America. When the Bering land bridge formed around 26 million years ago, one of the three-toed browsing horses, *Anchitherium*, crossed to Asia where it continued to live unchanged for several million years.

Many of the detailed changes in horse anatomy over 64 million years of evolution are associated with their increase in size.

Feet and legs As horses became larger, their legs grew longer in proportion. The bones between ankles and toes lengthened and they began to run on the tips of their toes. Even in the earliest horse, *Hyracotherium*, the third toe was the strongest and bore most of the animal's weight. However, it was not until they were living on the firm ground of the open grasslands that *Pliohippus*, the first one-toed horse, appeared.

Teeth The larger horses needed more food and they also changed from eating leaves to eating grass. Their teeth gradually became more complex to cope with the changed diet. Here a cheek tooth from each horse is shown after it has been worn down by some years of feeding. The teeth have strong ridges between rounded cusps and when the cusps wear down, the ridges remain, forming a complex pattern of sharp enamel.

	EURASIA	NORTH AMERICA	SOUTH AMERICA
HOLOCENE 11,000	Equus	Equus	Equus
PLEISTOCENE 2 m	Hipparion		Hippidion
PLIOCENE 7 m		Hipparion / Pliohippus	
MIOCENE 26 m	Anchitherium	Anchitherium	Merychippus
OLIGOCENE 38 m			Mesohippus
EOCENE 54 m	Hyracotherium		Hyracotherium

years ago

→ migration
green = plant eater

The horses – evolution and migration

At the beginning of the Miocene, when the grasslands spread so widely, a much larger horse, *Merychippus*, evolved in North America. *Merychippus* was the first grass-eating horse and its teeth were more complex, an adaptation to its new type of food. It was probably about 1m high at the shoulder.

Larger animals may find it difficult to reach food on the ground and *Merychippus* as a grazer had a longer neck than its browsing ancestors. Its head was supported by a ligament which stretched from the back of its skull to its shoulder. The ligament was made of elastin, a substance which stretched when the horse lowered its head and contracted again when it raised it. In this way it used the minimum amount of energy.

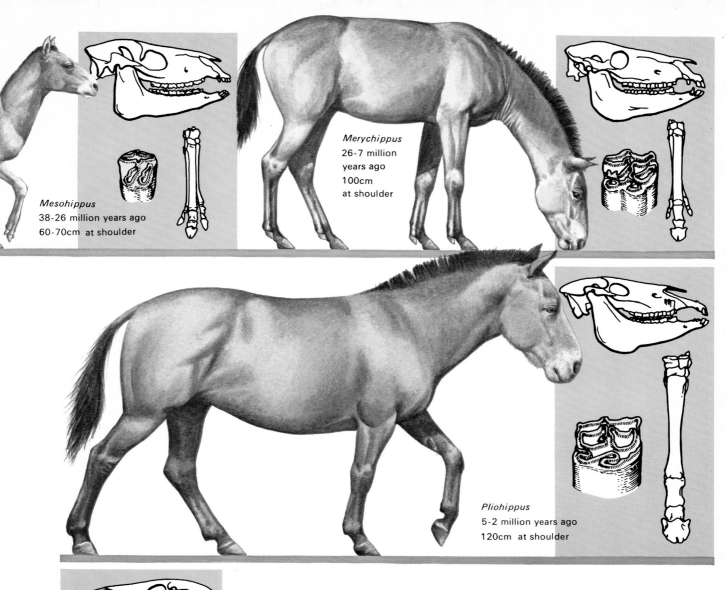

Mesohippus
38-26 million years ago
60-70cm at shoulder

Merychippus
26-7 million
years ago
100cm
at shoulder

Pliohippus
5-2 million years ago
120cm at shoulder

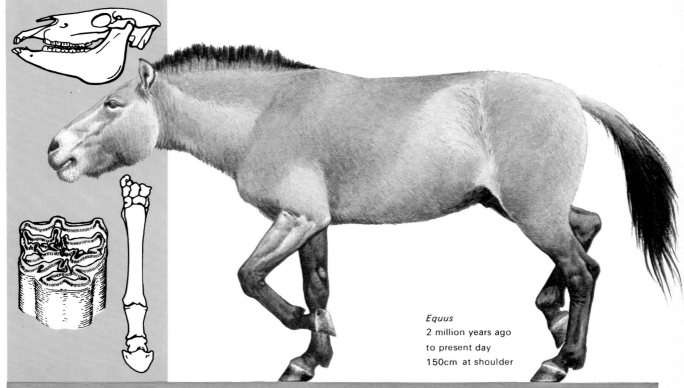

Equus
2 million years ago
to present day
150cm at shoulder

Horses did not need to be fast moving to obtain their food, but speed was essential if they were to avoid carnivores, especially in open woodland and grassland where sabre-tooth cats, wolves and lions were always ready to attack. As the horse moved into open country, the bones between its ankles and toes grew very long and it began to walk on the tips of its toes. The limbs from the knee to the elbow joints also lengthened but the upper limbs shortened. The major muscles controlling movement became concentrated near the shoulder and hip joints. This arrangement is a specialization for speed. A small contraction of the muscles moves the limb through a very wide arc so that the foot at the far end moves extremely fast.

Many different species of *Merychippus* were alive during the Pliocene, all managing to survive on the lush grasslands. As time passed, they divided into two main groups: one was to lead on to one-toed modern horses; the other, which included a horse called *Hipparion*, survived for over 5 million years but never evolved beyond the three-toed stage. *Hipparion* spread into Asia via the Bering land bridge and gradually made its way west to Europe. It was still living there at the beginning of the Pleistocene, 2 million years ago, but then died out.

The true horse line continued to evolve and live only in North America and there, some 5 million years ago, the first one-toed horse, *Pliohippus*, appeared. *Pliohippus* was very like a modern horse but smaller. The ability to run on tip-toe only developed once horses lived entirely on open grasslands, that is on firm ground. When the ground underfoot was soft, it was important for the toes to splay out to absorb the shock of its unevenness. Some of these horses managed to cross the heavily forested link between North and South America and there a new type evolved, *Hippidion*. *Hippidion* died out early in the Pleistocene.

At the very beginning of the Pleistocene, horses belonging to the modern genus *Equus*, evolved from *Pliohippus* in North America.

The first modern horses probably looked like Przewalski's horse, the wild horses that lived until recently in parts of Mongolia. Today a small herd is alive in Poland and there are several animals in zoos. They are stocky animals with a short, heavy head and a mane that stands up along their neck like a brush.

Equus was and is a very successful and adaptable animal. It spread into South America and Eurasia, taking the places of the earlier horses that had died out there. As the climate grew colder and the Ice Ages began, it spread south to Africa. In Asia and India wild asses evolved and in Africa, the zebras. Both groups still

Moropus

Miotapirus

have brush-like manes. The wild ass's sandy-coloured coat blends well with the semi-desert countryside where it lives while the zebra's stripes are an adaptation to the taller grass of the savannah. Seen from close to they are conspicuous but from a distance they blend together and provide an effective camouflage.

Curiously, about 15,000 years ago the horse died out in South America and, even more surprisingly, in its evolutionary home, North America. It was not until Europeans invaded South America in the sixteenth century that horses were reintroduced there. The wild mustangs of the North American prairies are descendants of horses abandoned by early Spanish settlers.

THE RHINOCEROSES

Many types of rhinoceroses had evolved in the Eocene and Oligocene in both Eurasia and North America. Some new forms appeared in the Miocene but from this

By the Miocene the only surviving odd-toed ungulates were horses, rhinoceroses, tapirs and chalicotheres. *Trigonias* and *Miotapirus* were direct ancestors of modern rhinos and tapirs and became established in their modern niches. The strange, clawed chalicotheres with their horse-like heads, strong front limbs and sloping backs, had first evolved in the Eocene. *Moropus*, the size of a large horse, is typical of the later animals.

The odd-toed ungulates

Trigonias

time onwards, the group was declining rather than becoming more varied.

The giant hornless rhinoceroses such as *Baluchitherium*, which had dominated the Asian landscape, became extinct in the Miocene and during the Pliocene all the rhinoceroses disappeared from North America. In Europe and Asia, however, they continued to flourish. The Sumatran rhinoceros, which had evolved in the Oligocene and still lives in Sumatra, was living all over Europe and Asia at this time. From it the woolly rhino evolved in the Pleistocene.

The main evolutionary line, descended from the hornless *Trigonias* of the Oligocene, continued to develop and, late in the Miocene, about 10 million years ago, modern Asian rhinoceroses appeared. Unlike the Sumatran rhinoceros they have only one horn.

During the Miocene the rhinoceroses crossed over into Africa where two species are still living today. The black rhinoceros is the more common, living in open country and browsing off the leaves of shrubs and trees. The square-lipped or white rhinoceros is a grazer and can crop grasses very close to the ground. It is now the second largest land mammal – only elephants are bigger – but unfortunately it may be on the verge of extinction.

CHALICOTHERES AND TAPIRS

The two other groups of odd-toed ungulates still alive in the Miocene were the chalicotheres and the tapirs. The strange, clawed chalicotheres lived in Europe until the end of the Pliocene and survived even longer in Africa, India and China. Tapirs were almost unchanged from their Eocene ancestors and have changed little since. The first tapirs belonging to their modern genus evolved in the Pliocene, the main difference being that they had increased in size. Even today they look very like their earliest known forebears.

63

Gomphotherium

Platybelodon

Phiomia

Moeritherium

Anancus

Mammuthus

The elephants

Moeritherium, the first of the elephant family, lived
in Africa during the Eocene and Oligocene epochs.
About the size of a pigmy hippopotamus, it was
probably a swamp-living plant eater.

Like other animal groups, elephants tended to increase in size as
they evolved. Their legs became strong to support their increased
weight and their heads were further from the ground. In this situation
some animals (such as horses) grew longer necks. Elephants' necks
actually became shorter to support their heavy heads but their lower
jaws grew long and they developed trunks so that they could reach
their food.

Phiomia and *Palaeomastodon* both evolved from *Moeritherium in*
the Oligocene. Each founded a separate evolutionary line. During the
Miocene 'four tuskers' such as *Gomphotherium* evolved from *Phiomia.*
Their upper tusks grew from bones at the front of their skull; their
lower were the long front teeth of their lower jaw. From the four
tuskers a group known as shovel tuskers evolved. These first appeared
towards the end of the Miocene in Asia but developed several new
forms in North America during the Pliocene. In one genus, *Platybelodon*, the lower jaw grew into a flat scoop or shovel, the wide
teeth at the tip forming a strong blade. These elephants shovelled up
food using their snout to scoop it into their mouth – just like a brush
and shovel.

The line from *Phiomia* ended with the 4m high European mastodon
Anancus which only died out around 10,000 years ago. By the time
Anancus evolved, the trunk was more important for food gathering
than the front teeth and these, including the lower tusks, had disappeared. The lower jaw grew shorter and the trunk, no longer supported,
hung down. The two upper tusks remained. In *Anancus* they were
straight weapons some 3 to 4m long.

Palaeomastodon, the larger of the two Oligocene elephants, was
the ancestor of *Deinotherium*, the hairy American *Mastodon* and
Stegodon. *Deinotherium* first appeared early in the Miocene in
Africa. Unlike all the other elephants it had no upper tusks though it
still had long, tusk-like lower front teeth. The part of the jaw where
these grew was bent over towards the ground and the downward
pointing tusks may have been used for digging. *Deinotherium*
spread to Europe and Asia and some species grew to 4m high. The
American *Mastodon*, a forest browser, survived until around 8,000
years ago.

Stegodon, which evolved in Asia late in the Pliocene and died out
during the Ice Ages, was the immediate ancestor of the mammoths and
of the two living elephants the African *Loxodonta* and the Indian
Elephas. The mammoths were the most successful of all Ice Age
mammals. In the north, they grew thick coats of reddish hair to protect
them from the extreme cold and spread all over Europe, Asia and into
North America. In warmer areas mammoths such as the North American imperial mammoth flourished and even migrated to South
America. They survived there as late as 600 A.D.

Today's Indian elephants may be closely related to the extinct
mammoths. African elephants descended from *Stegodon* via the
straight-tusked *Palaeoloxodon* of the Pleistocene.

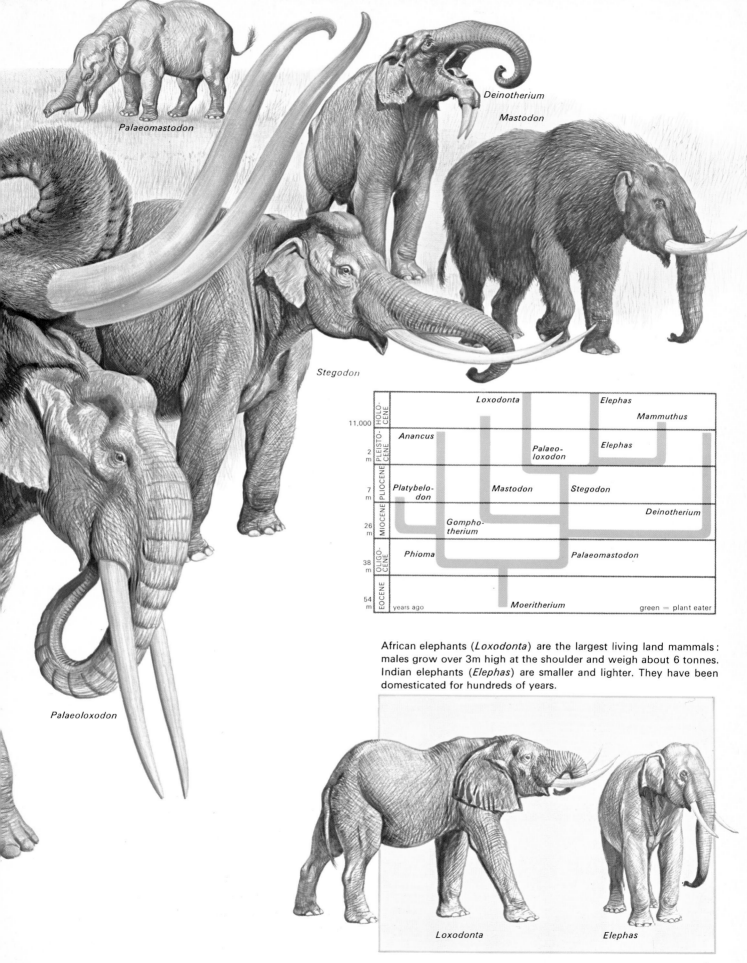

Palaeomastodon

Deinotherium

Mastodon

Stegodon

Palaeoloxodon

		Loxodonta		Elephas	
HOLOCENE 11,000					Mammuthus
PLEISTOCENE 2 m	Anancus		Palaeo-loxodon	Elephas	
PLIOCENE 7 m	Platybelo-don	Mastodon	Stegodon		
MIOCENE 26 m		Gompho-therium			Deinotherium
OLIGOCENE 38 m	Phioma			Palaeomastodon	
EOCENE 54 m	years ago		Moeritherium	green = plant eater	

African elephants (*Loxodonta*) are the largest living land mammals: males grow over 3m high at the shoulder and weigh about 6 tonnes. Indian elephants (*Elephas*) are smaller and lighter. They have been domesticated for hundreds of years.

Loxodonta

Elephas

The even-toed ungulates

The even-toed ungulates were at their most successful during the Miocene and Pliocene epochs and in number and variety must have been the dominant plant eaters.

The first group – the pigs in Eurasia and the peccaries in North America – had evolved by the Oligocene (38 million years ago). During the Miocene a large number of different types of pigs appeared and it was at this time that they moved into Africa. One genus from the Caucasus, *Kubanochoerus*, had a bony horn on its forehead but it was the only pig ever to develop one and most relied on their tusk-like front teeth to protect them from predators. Today's European wild boar, African warthog and Asian babirusa are well able to defend themselves.

Peccaries never crossed into Europe and Asia but in the Miocene they managed to migrate across the densely forested land bridge to South America. As forest animals they were forced to move when the spreading grasslands destroyed their usual environment. They are well adapted to forest life: they feed on leaves, fruit, acorns, cacti and roots and with their relatively long legs can move through the undergrowth at high speed. In the Pliocene there were several large peccaries, leading to the flat-headed and long-snouted peccaries of the Pleistocene. Both these large animals died out 10,000 years ago. Smaller peccaries still live in South America but have disappeared from most of North America.

Another group of pigs, the anthracotheres, were living in Europe, Asia, Africa and North America at the end of the Oligocene. Anthracotheres lived like hippopotamuses in the lakes and rivers for millions of years but during the Miocene they died out in North America, Europe and Asia. They also began to decline in Africa but there, in the Pliocene, true hippopotamuses evolved from them.

Hippopotamuses spread from Africa into Europe and Asia but today are once more found only in Africa. In Europe they died out towards the end of the Pleistocene, just over 11,000 years ago; the last Asian hippopotamuses became extinct in India and the East Indies in the last few hundred years.

The ruminants

Pigs and peccaries form the first group of even-toed ungulates. The second is the ruminants, a very large group which includes deer, antelopes, camels, sheep, goats, cattle and many more. They were without doubt the most important animals of the Miocene and Pliocene epochs. As the grasslands spread over large areas of North America and Eurasia, many of the plant eaters

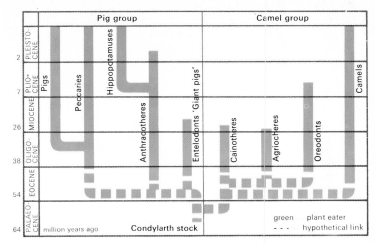

The even-toed ungulates   Pig and camel groups

Bothriodon

became adapted to eating the tough grasses and grazing instead of browsing became the main way of life for many herbivores. The animals that adapted best to the new diet were the ruminants. From their fossil remains alone, and particularly from their teeth, it is hard to imagine that the ruminants were more efficient plant eaters than, for example, the horses. Their teeth seem much less well adapted for dealing with the tough grasses they fed on. Their success was, in fact, due to their highly specialized digestive system which broke down cellulose and extracted the protein.

The first known ancestors of the ruminants had evolved in the Eocene and the forerunners of camels, giraffes and deer were alive in the Oligocene. Now the ruminants began an explosive development.

Camels and llamas are the most primitive types of ruminants still living. Their evolution took place entirely in North America. *Poebrotherium*, their common ancestor, evolved there in the Oligocene and by the Miocene all the basic features of a modern camel

Hippopotamus

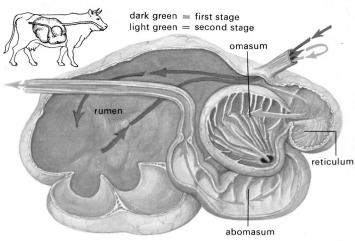

dark green = first stage
light green = second stage

omasum

rumen

reticulum

abomasum

A ruminant's stomach is divided into four major sections. The largest part is the rumen, where food goes first after it has been swallowed. In the rumen the food is mixed with mucus and broken down into sugars and various organic acids by bacteria. Small stones and extra tough material are collected in the reticulum. When the plant material is a pulp, the animal brings it back into its mouth where it is pounded and ground into a fibrous ball. The ball of food is swallowed once again but this time it goes to another part of the stomach, the omasum. Here it is squeezed until all the water is removed. The final stage takes place in the abomasum or true stomach, where acid enzymes digest the food and extract the protein.

Throughout the Age of Mammals different groups of animals had occupied the niche finally taken over by true hippopotamuses in the Pliocene. By far the best adapted animals for this way of life, they evolved in Africa from anthracotheres such as *Bothriodon*. Structurally they were almost exactly like modern hippos; the main difference was that their eyes were even further towards the top of their head.

skeleton had developed. They became a very varied group, filling several ecological niches that in Europe and Asia were occupied by other types of mammals. There was a 3m tall giraffe-like camel, *Alticamelus*, with an exaggeratedly long neck, specialized for feeding off the leaves from the tops of trees – exactly like a modern giraffe. There were also a number of other long-legged, long-necked types and a minute gazelle-like camel which filled the niche of Eurasian gazelles.

During the Pliocene llamas reached South America and camels Asia. By the Pleistocene they had moved into eastern Europe and North Africa. They became completely extinct in North America towards the end of the Pleistocene and today they survive only in the deserts of Asia and Africa. Llamas and their relatives the guanacos, alpacas and vicunas, live only in the mountains and colder regions of South America.

Although as far as their stomachs are concerned they are more primitive ruminants, camels are highly adapted for the difficult conditions in which they live.

Next page: As the grasslands spread in the Miocene, more and more even-toed ungulates evolved. In North America there were several horned traguloids such as *Synthetoceras* (1) and *Syndyoceras* (2); hornless forerunners of deer and giraffes (*Blastomeryx* 3); and several camels. *Oxydactylus* (4) was a grazing camel but *Alticamelus* (5) had evolved a long giraffe-like neck and fed on the leaves of trees. The only successful odd-toed ungulates were the horses (*Merychippus* 6) and rhinoceroses such as *Teloceros* (7) and *Diceratherium* (8).

The evolution of the pecorans

Pecorans is the name given to the large group of ruminants whose many-chambered stomachs are most perfectly adapted to grass eating. They include all the familiar grassland deer and antelopes as well as domestic animals such as cattle, sheep and goats. The pecorans are divided into three groups, the traguloids, cervoids and bovoids. Their ancestors lived in both Eurasia and North America but during the 24 million years of the Miocene and Pliocene they evolved separately on each continent.

THE TRAGULOIDS

The traguloids are the most primitive group of the pecorans and first appeared in the Eocene in Mongolia over 40 million years ago. Early traguloids were small, some only as big as rabbits and these, though common in the Oligocene, died out early in the Miocene. A second group, the chevrotains, evolved from them. The first traguloid to show the beginnings of horns was *Protoceras* and from this a third group of rather larger animals arose. Their horns, covered in skin, were probably not shed.

The horns served two purposes: they were useful in defence and especially in trials of strength during the breeding season when the strongest males fought for the females of the herd. They were also recognition signals from a distance, ensuring that males and females of the same species could get together to ensure the production of fertile young.

Today the only remaining members of this once large group are the chevrotains, small hornless animals that are almost exactly like one of the earliest traguloids from the Oligocene. Instead of horns they have long, stabbing canine teeth which hang outside their mouths and can be used as weapons of defence. They have three- not four-chambered stomachs and the third and fourth toes remain as small dew claws at the side of their legs.

THE CERVOIDS

The second major division of the pecorans is the cervoids, the deer and giraffes. Most of these are browsers but all have four-chambered stomachs capable of dealing with tough plant food. They are thought to be descended from an early traguloid.

The first cervoid was an animal named *Blastomeryx* which lived in North America and Eurasia during the Miocene. *Blastomeryx* itself was hornless but some of its descendants had long, skin-covered horns ending in a tiny, two-branched antler. It was from these first horned cervoids that today's deer and giraffes evolved, as well as a group of more primitive deer, the palaeomerycids.

Deer have a pair of branching antlers which grow

Animals like the North American *Cranioceras* were descendants of hornless traguloids and were themselves the immediate ancestors of both deer and giraffes. *Dicrocerus* was the first true deer; *Eucladocerus*, a later species, had the most complex antlers of all.

from a short bony stump and are shed every year. Each year the antler part adds an extra branch until the maximum number for the particular species is reached.

Although antlers look effective weapons, they are rarely used for fighting except in the mating season when the males lock antlers, pushing and shoving in a ritual contest for the females of the herd.

Large antlers seem a disadvantage in forest and woodland and you might expect that natural selection would discourage them. However, it is the strongest males with the largest antlers that win the fight for the females and so mate with them. Their magnificent head decorations are thus passed on to a new generation.

The first true deer, *Dicrocerus*, appeared in the Miocene and by the Pliocene a large number of deer had evolved, each with different types of antlers. Several living genera were there, including the red deer, fallow and roe deer.

For some unknown reason, neither deer nor giraffes evolved in North America, although their ancestor *Blastomeryx* lived there and conditions were generally very much the same. The primitive palaeomerycids did evolve there, many developing very complex antlers and

...oceras

Sivatherium

Palaeotragus

The pecorans

	Traguloids		Cervoids		Bovoids	
PLEISTO-CENE	Chevrotains		Deer	Giraffes	Pronghorns	Bovids
PLIOCENE	Protoceratids		Palaeo-merycids			
MIOCENE	Hypertragulids					
OLIGO-CENE						
EOCENE	million years ago	Condylarth stock	green plant eater - - - hypothetical link			

Early giraffes were divided into two groups. One included *Palaeotragus*, the ancestor of modern giraffes and okapis. The other led to *Sivatherium*, a 2.5m tall giant which became extinct about 2 million years ago.

horns. They continued to flourish throughout the Pliocene. Then, in the Pleistocene, deer from Eurasia reached North America, and the North American forms were unable to compete. The invading Eurasian deer soon settled in their adopted home and familiar forms such as the wapiti or elk and the moose evolved. Later, in the Pleistocene, there were giant elks and mooses; the giant Irish elk's antlers were up to 3.5m across.

The deer invaded Africa in the Pliocene but only managed to reach the northern fringes of the continent. During the Pleistocene they also migrated into South America. One genus *Rangifer*, the reindeer of Eurasia and the caribou of North America, became adapted to life on the tundra of the cold arctic and remained there, migrating southwards in the winter to find food.

The second line from the early cervoids lead to the giraffes and okapis which evolved in Eurasia and never crossed to North America. Their place there was filled by the giraffe-like camels. Giraffes' legs are very highly developed and they no longer have the remains of their third and fourth toes as dew claws. Their horns, however, are like those of earlier cervoids. Today's giraffe has two bony projections on the top of its head, covered in skin with a tuft of hair at the tips. They are the equivalent of the early stages of antler growth when there is just a simple spike of bone enclosed in velvet. However, unlike antlers they never grow any larger and they are never shed.

The first record of the giraffes comes from the Miocene in India where *Giraffokeryx* was similar to modern okapis. It was a forest browser and already had a long neck to enable it to reach the higher branches of the trees. Modern giraffes and okapis are probably descended from *Palaeotragus*, another okapi-like animal which evolved in the Pliocene. Giraffes seem to have reached Africa during the Pleistocene and are now found only there. Their enormously long necks and legs are an adaptation to the complex web of life on the grasslands where, to survive, each type of plant eater must exploit a specialized niche.

During the Miocene a second branch of the giraffe family appeared, the sivatheres. The first of these was *Prolibytherium*, from the Miocene in North Africa. It was about the size of a sheep, with a pair of large fan-shaped horns. The last of the line was the giant *Sivatherium*, which evolved in India during the Pleistocene. *Sivatherium* stood about 2.5m at the shoulder and had massive antler-like head ornaments which were never shed. With its powerful shoulder muscles and long, strong front legs, it was probably the Indian equivalent of the giant Irish elk.

THE BOVOIDS

Bovoids, the most efficient of all grass eaters, evolved in both Eurasia and North America. Like the cervoids, they first appeared in the Miocene and were probably descended from one of the small traguloids. The five Eurasian groups are cattle (including spiral horned antelopes), horse antelopes (including kobs, waterbuck, hartebeests, oryxes and gnus), true antelopes (including small gazelles), duikers, sheep and goats. The North American bovoids are pronghorns, now a single species.

Bovoids have a pair of simple, unbranched horns made up of a bony core covered by a horny sheath. Except for the North American pronghorn, this sheath is not shed. Sometimes the horns grow long, with graceful backward curves; at other times they curve out sidewards – as in cattle. Whatever the shape they are really effective weapons and can gore predators to death or impale them on their points. They, like deers'

Horns and antlers
Antlers are shed every year. The most impressive of all belonged to the giant Irish elk *Megaceros* and spanned more than 3m. Its modern relatives the red deer have basically similar antlers but of more manageable size. Usually only male animals have antlers.

The most primitive type of head ornament is a bony outgrowth covered in skin. *Prolibyotherium*, an ancestor of the giraffe, probably sloughed off this skin every year. Today's giraffes have small, bony skin-covered knobs but do not shed the skin.

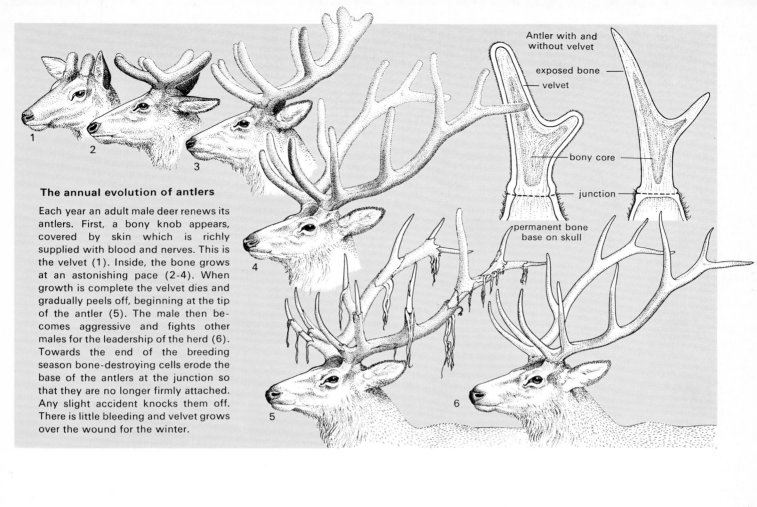

The annual evolution of antlers

Each year an adult male deer renews its antlers. First, a bony knob appears, covered by skin which is richly supplied with blood and nerves. This is the velvet (1). Inside, the bone grows at an astonishing pace (2-4). When growth is complete the velvet dies and gradually peels off, beginning at the tip of the antler (5). The male then becomes aggressive and fights other males for the leadership of the herd (6). Towards the end of the breeding season bone-destroying cells erode the base of the antlers at the junction so that they are no longer firmly attached. Any slight accident knocks them off. There is little bleeding and velvet grows over the wound for the winter.

Antler with and without velvet
exposed bone
velvet
bony core
junction
permanent bone base on skull

Left: The aurochs, ancestor of modern European cattle, had true horns. These are made up of the fibrous protein keratin and have a hard, bony core. They grow like nails until they reach their full size but are not shed. Both males and females carry them.

Synthetoceras, one of the very varied North American ruminants of the Miocene-Pliocene, had divided horns, not antlers. Modern pronghorns also have divided horns and shed the outer covering annually.

The most primitive ruminants, such as *Archaeomeryx*, had neither horns nor antlers. Instead they used long upper canine teeth as a means of defence. The tiny modern chevrotain, little larger than a hare, has the same primitive features.

antlers, are used as reliable recognition symbols.

The major evolutionary development of the bovoids began in Eurasia in the Miocene with the small woodland antelope *Eotragus* and the grazing gazelle *Gazella* which is still alive today. These two small animals gave no hint of the explosive evolution of the group that was to take place in the succeeding Pliocene. In Asia over 50 genera have been recorded, with about 20 in Europe and 12 in Africa.

Most of our modern domestic animals are descended from the bovoids which evolved in Europe and Asia during the Miocene and Pliocene.

In the Pliocene the Asian grasslands were full of vast roaming herds of animals that are today familiar inhabitants of the savannahs of East, West and Southern Africa. There were duikers and many types of small antelopes while among the cattle group were the ancestors of the eland and the kudu. The nilgai, the ancestor of true cattle, was also there. It did not die out when cattle later evolved, but survives today in India.

The direct ancestors of modern types of cattle first appeared in India in the Pleistocene. The lightly built ox *Leptobos* was built like an antelope but its horns turned forwards as well as sideways. Three main lines developed in the Pleistocene from *Leptobos*: the Asian water buffalo, the aurochs and bison group and the African buffaloes.

The aurochs were the wild cattle of the Pleistocene and it is from them that domestic cattle were bred. They were large animals, standing 2m high at the shoulder with fierce, curving horns. The last was killed by man in Poland in 1627 but long before that they had been domesticated and their descendants survive now in many parts of the world. Bison, which also originated in India, spread both east and west. The European species became extinct in the wild in 1921 but survivors in zoos were reintroduced into Polish forests and today there is a herd of around a thousand. With its thick coat the bison was able to cross to North America during the Pleistocene and there it rapidly spread over all the plains, reaching as far south as El Salvador.

The African buffaloes, whose horns meet across their heads at the base, are the only survivors of the third evolutionary line from *Leptobos*.

Horse antelopes also evolved on the Asian grasslands in the Pliocene. An early oryx, *Paleoreas*, flourished in herds in Europe, Asia and Africa though it is now found only in the drier parts of Africa and the Middle East. Also in the Pliocene were ancestors of the roan and sable antelopes which, with their long ears, general horse-like shape and backward sloping horns, gave this

The most advanced of all the ruminants and by far the most successful of all plant eaters are the bovoids. By the Pliocene, 7 million years ago, all the main groups had evolved, many already in their modern forms. The nilgai (probable ancestor of cattle) duiker and chamois (ancestor of sheep and goats) are all alive today. *Palaeoreas* is an early oryx and *Gazella* is an early form of gazelle.

Nilgai

Duiker

group the descriptive name of horse antelopes.

Both true antelopes and horse antelopes were adapted to life in warm temperate woodland and savannah grasslands and as the climate grew worse in the Pleistocene, most died out in Europe and migrated southwards. Today they are concentrated in India, tropical Asia and Africa, where they seem to be at the peak of their evolutionary development, with many different species living side by side.

Sheep and goats both evolved in the Pliocene but were adapted to upland areas and could stand greater extremes of temperature than the cattle and antelopes.

The primitive members of the group include the living chamois or mountain goat from which true sheep and goats evolved. One of the first of its descendants to become adapted to life in the cold desert regions of Asia was the saiga which was living there in Pliocene times. The saiga's large, fleshy snout encloses a large sac lined with mucous membrane. This warms the air and also filters fine dust from it – a special adaptation to cold desert conditions.

Gazella

Chamois

Palaeoreas

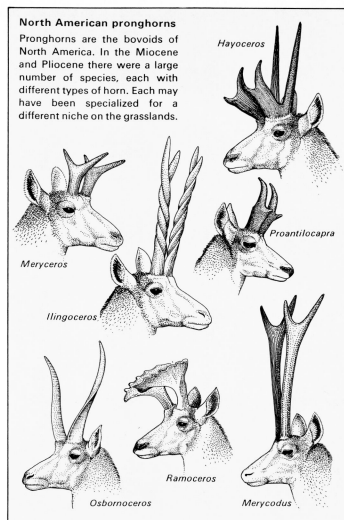

North American pronghorns
Pronghorns are the bovoids of North America. In the Miocene and Pliocene there were a large number of species, each with different types of horn. Each may have been specialized for a different niche on the grasslands.

Hayoceros

Meryceros

Ilingoceros

Proantilocapra

Osbornoceros

Ramoceros

Merycodus

During the Pleistocene when tundra spread over large parts of Europe and Asia, the saiga migrated westwards into Europe and also across the Bering land bridge into Alaska. Today it is only found in central Asia to the north of the Himalayas.

A second cold-climate member of the sheep and goat group evolved in the Pliocene – the musk-ox. In the middle of the Pleistocene, when the land link with North America was well within the cold Arctic region, it was one of the few animals to migrate across. Today it is established in Arctic regions of Canada and Greenland, but has disappeared from Eurasia. The takin which lives in the mountains of Tibet may be a surviving descendant of the Eurasian musk-oxen.

Antelopes are very specialized in diet, to reduce competition among the different groups. Sheep and, even more so, goats, are extremely unspecialized and can eat anything that is even vaguely vegetable. Goats are agile enough to climb trees as well as scale apparently impossible slopes in search of sparse grasses. In competition they will survive at the expense of everything else.

During the Pleistocene cattle, sheep and goats were to become the most important mammals on Earth – except, perhaps for man.

BOVOIDS IN NORTH AMERICA
The pronghorns were the only bovoids to evolve in North America. Their immediate ancestor *Proantilocapra* was living in the Pliocene and at that time there were also a very large number of other pronghorn species which have left no descendants. Pronghorns have horns which never lose their bony core but shed the branching horny sheath every year. In the Miocene and Pliocene there were pronghorns with gracefully curving horns like African waterbuck, others with several simple horns, or two spirally twisted ones. Another group had horns with three or more branches. Like the many different grass eaters of the African savannah today, they probably occupied different niches within the grassland environment, feeding off different levels and types of vegetation. Unfortunately there can never be any way of finding out the exact details.

The flesh eaters

By the end of the Oligocene the two main groups of modern carnivores had already separated into their distinct evolutionary lines. The first of the cat family, the mongooses and civets were just beginning to appear and there were early members of the dog family such as weasels and polecats. As the forests and woodlands thinned out and grassland animals became the most common plant eaters, the flesh eaters also had to change and adapt to life on the open plains.

THE EVOLUTION OF THE CATS

Most of the animals grazing on the grasslands were swift runners, with an acute sense of sight and smell. Some, such as the bovoids, were well armed with horns that could easily kill an unwary predator. Because the countryside was open, it was difficult to approach a grazing animal without being noticed and in order to succeed, the predators needed to be highly intelligent. It is because cats *were* intelligent that they were such successful hunters.

Cats hunt primarily by stealth. With their soft padded feet and retractable claws they can stalk their prey and, when they are within striking distance, make a short burst of speed to pounce, knocking their victim to the ground. With a rapid bite, they can sever its spinal cord and neck arteries. Cats' teeth are specially adapted for cutting flesh and their jaws enable them to hold on tightly, even to a struggling animal.

For most of their history, cats can be divided into two contrasting groups, the sabre-tooth stabbing cats and the biting cats. Both these had begun to evolve at the end of the Oligocene, soon after the long-bodied civet types. While primitive cats remain small, woodland hunters to the present day, the two other groups continued to adapt and change.

The ancestor of both the sabre-tooths and the biting cats was *Dinictis*, which evolved in North America some time in the Oligocene. Their upper canine teeth were much larger than those of modern cats but shorter than those of typical sabre-tooths. They are sometimes called 'false sabre-tooths'.

The sabre-tooth line consisted of cats with increasingly large stabbing front canines: *Eusmilus*, which evolved in the Oligocene had such long teeth that its lower jaw had to open to an angle of 90° before it could effectively stab its prey. Although its teeth were long, its jaw muscles were quite weak but it had very powerful neck muscles. Sabre-tooths were not specially fast runners and probably hunted large, slow-moving mammals that would die from loss of blood once their arteries had been cut;

there was no need for the hunter to hang on as the animal struggled to run away.

In the Pliocene the sabre-tooth *Machairodus* lived all over Europe, Asia, Africa and North America. It died out in the Old World by the middle of the Pleistocene but survived in North America until near the end of the period. Another, *Smilodon*, was probably the most successful of all sabre-tooths, but was also the end of its evolutionary line. It lived all through the Pleistocene and spread widely over the world, even reaching South America. About 11,000 years ago all the true sabre-tooths became extinct.

Towards the end of the Pliocene a new type of sabre-tooth evolved, *Homotherium*. This is sometimes known as a scimitar-toothed cat because its canines were shorter, but razor sharp. Its most unusual feature was its long front legs and short, bear-like back legs which made it look almost as if it were walking upright. Scimitar cats survived in Europe and North America until about 20,000 years ago. They may have been specially adapted to preying on mammoths for in a cave in Texas a number of adults and cubs were found with the bones of young mammoths.

The evolution of the biting cats ran parallel to the sabre-tooth cats. An early side-branch was a cat named *Nimravus* which was more long-limbed than the other types and seems to have been specialized for running, like a cheetah. The main evolutionary line led to *Pseudaelurus* which evolved in Africa during the Miocene. *Pseudaelurus* was about the size and proportions of a modern leopard and was very similar to today's species. It spread to Europe and Asia and by the early Pliocene, 8 million years ago, cats of the modern genus *Felis* had evolved.

The first fossil *Felis* was an animal the size of a domestic tabby cat. By the Pleistocene various species had reached all parts of the world, including Africa and South America. These small cats were probably successful because of the sudden increase in numbers of rats and mice that occurred about the same time. The larger cats such as pumas, lions, leopards, tigers and cave lions are also included in the *Felis* genus and evolved at around the same time.

In the Old World the cheetah, a highly specialized runner, evolved in the Pleistocene. The fastest of all mammals, it can run at speeds up to 120km/h. Cheetahs never reached North America but there a relation of the puma or mountain lion developed in the same way. *Felis trumani* must have filled the same ecological niche, preying on the fastest of the deer and pronghorns.

When the sabre-tooth cats died out at the end of the

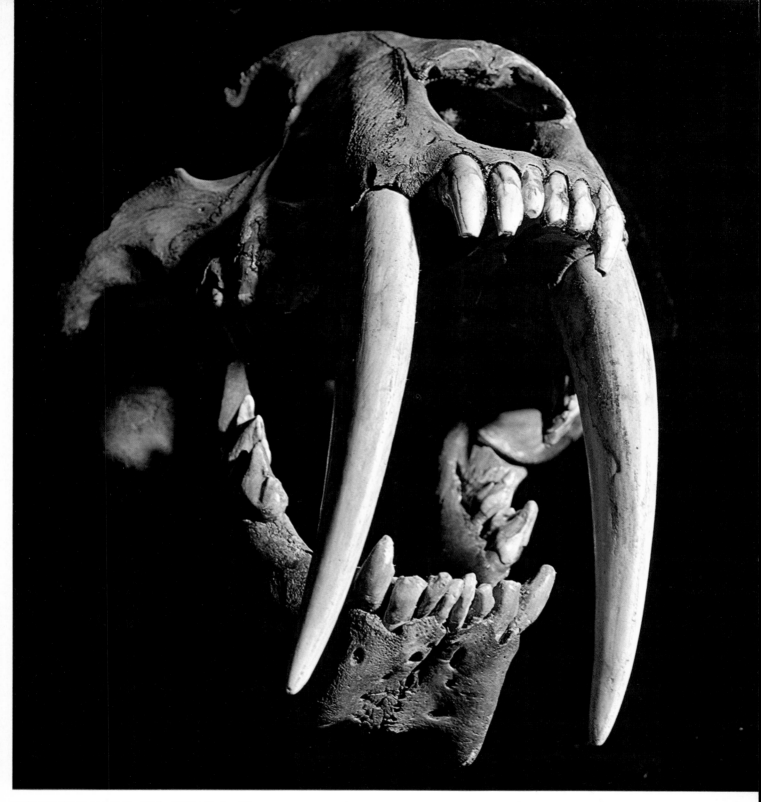

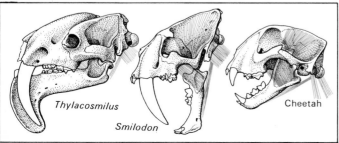

Thylacosmilus

Smilodon

Cheetah

Sabre-tooth cats such as *Smilodon* (above and centre left) had long, dagger-like canine teeth and powerful neck muscles which pulled their head downwards with great force, driving their great teeth into their prey. Their lower jaw and the muscles for closing it were weak: sabre-tooths could not hold on for long to a struggling animal. In contrast, the biting cats (left) have shorter canines, strong lower jaws and powerful jaw-closing muscles. Once attached to their prey, they hang on until the death. The marsupial sabre-tooth *Thylacosmilus* (far left) was not related to placental sabre-tooth cats but had evolved the same kind of teeth and a similar muscle structure.

Dinictis

Both stabbing and biting cats evolved from a common ancestor, *Dinictis*. The stabbing line ended with *Smilodon*, the 'sabre-tooth tiger' of the Pleistocene.

Eusmilus

Hoplophoneus

STABBING CATS

Pseudaelurus

Felis

BITING CATS

Nimravus

Pleistocene, the biting cats continued to thrive. Many of the large plant eaters that formed the sabre-tooths' prey became extinct, but there were still a great variety of smaller, more agile ungulates. The biting cats, being faster than the sabre-tooths, were able to hunt these successfully.

Today there are 36 species of cats, including the domestic cat *Felis catus*. Selective breeding has produced a very great number of different 'breeds' of domestic cats but as they can all mate and produce fertile young, they are still included in a single species.

VIVERRIDS, HYAENAS AND AARDWOLVES

The primitive cats from which the true cats originated are the viverrids, the small, long-bodied forest hunters that had changed little from their Eocene ancestors, the miacids. Modern mongooses and civets evolved first in Europe but by the Pliocene had spread to Asia and by the Pleistocene, to Africa. They cannot survive cold climates and are now found only in the warmer parts of Africa and Eurasia.

Hyaenas developed from the viverrids in Europe. The spotted hyaena and the striped hyaena both appeared in the Pliocene and spread across Asia and into Africa. Hyaenodonts, the primitive hyaena-like hunters and scavengers that had for about 30 million years been Africa's only large flesh eaters finally died out as the true hyaenas and the sabre-tooths invaded

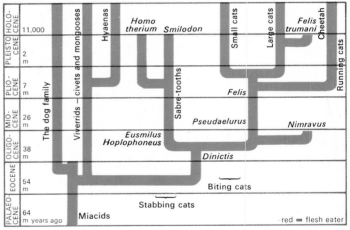

The flesh eaters – the cats

their territory and began to hunt their plant-eating prey.

Hyaenas are the most recent evolutionary development among the flesh eaters and now fill the role of general scavenger in the tropical grasslands. They can also hunt like dogs, in packs, and in North America their place was taken by a specialized group of dogs. Only one type of hyaena actually reached North America during the Pleistocene – a fast cheetah-like animal, *Chasmaporthetes*. Another specialized side-branch of the hyaena line is the aardwolf, which evolved in Africa in recent times. Aardwolves feed on termites and white ants – a very unusual diet for one of the cat group.

Homotherium

Smilodon

RUNNING CATS

Felis trumani

Cheetah

Several biting cats were specialized for speed. *Nimravus* was an early runner; the cheetah developed similar characteristics much later. In North America a species of puma *Felis trumani* (now extinct) took the Old World cheetah's place.

Cave lion

Tiger

Leopard

Lynx

European wild cat

Domestic cat

79

THE EVOLUTION OF THE DOGS

The earliest members of the dog family come from the Oligocene, when their evolutionary line split off from the more primitive miacid group. At first they continued to be long-bodied, short-legged animals such as *Pseudocynodictis* and *Mesocyon*, though there were some larger types. Then, in Europe during the Pliocene, the first of the modern dog genus, *Canis* evolved. This genus, which includes wolves, jackals, coyotes and dogs, soon spread to Asia and Africa. In the Pleistocene they reached North America and invaded South America. The fox originated in North America in the Pliocene and spread to Eurasia during the Pleistocene.

The evolution of the dogs shows a different way of hunting on the open plains of the Miocene grasslands. Again the key to their success is their high level of intelligence but unlike the cats, they also use teamwork.

The pack cuts out a member of a herd of plant eaters – usually a young or weak animal – and drives it to and fro until it gradually tires. They then close in and pull it to the ground. Physically dogs are specialized for fast running. They run on the tips of their toes, and their heart and muscles are adapted for maintaining a high speed. They are also known to have great powers of endurance. Once they have selected their prey, there is little chance of its escaping either by speed or cunning.

Their system of hunting is complicated and as no two hunts will be the same, dogs must be highly adaptable. Group hunting is only really successful if it is very carefully organized and a dog pack has a leader and a clearly defined heirarchy. Each member plays its particular part in the operation and shares the kill – again in order of seniority.

Although modern types of dog were established in Eurasia and Africa from Pliocene times, it was in North America that they became most varied.

During the Miocene the hyaena-like osteobores evolved as scavengers, to be replaced in the Pleistocene by heavily built dire wolves. The bear-dogs which had evolved at the end of the Oligocene also appeared first in North America where they filled the ecological niche of true bears. They spread to Asia, Africa and Europe before dying out in the Pliocene.

Today there are 37 species of true dogs. Domestic dogs, like domestic cats, are included in a single species.

BEARS AND RACCOONS

Two groups of omnivorous animals are included in the dog section of the carnivores – the bears and the raccoon family. Bears first appeared in Europe in the Miocene, probably descended from early dogs. By the end of the epoch, the modern form had evolved. During the Pleistocene they migrated to North America and a new type, the short-faced spectacled bear evolved. One species, the cave bear, survived for much of the Ice Age by hibernating in caves during the coldest parts of the winter.

Bears are omnivores, eating fruits, nuts, eggs, fish and even small mammals. They are therefore very adaptable and, being mainly quite large animals, they also have few enemies. Seven species are still living one, the polar bear, specially adapted for Arctic conditions and a diet of fish and seals. The most highly specialized of all bears is the giant panda, which has become a plant eater, eating only bamboo shoots.

The other group of omnivores is the raccoon family, which includes the lesser or red panda. The first raccoons evolved during the Oligocene and were from the first basically tree dwellers. Early fossils come from both North America and Eurasia and during the Pliocene they invaded South America. Today raccoons are found only in the New World.

Daphoenus

Osteobores

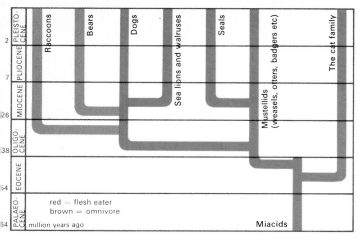

The flesh eaters – the dogs

red = flesh eater
brown = omnivore
million years ago

THE MUSTELLIDS

The most primitive members of the dog family, the mustellids, continued to develop along a separate evolutionary line. Today they are animals such as weasels, stoats, polecats and badgers, all of which are noted for their fearlessness and ferocity. They have long bodies and short, stocky limbs, looking generally very like the early miacids from which they evolved. Most live in trees in northern temperate countries although there are some in Africa and South America.

The earliest mustellids had been living in the Oligocene in Eurasia and by the Miocene there were animals such as weasels and polecats. The ancestor of the wolverine or glutton evolved during the Miocene. The wolverine is the largest living mustellid and though it feeds mainly on lemmings, it has been known to kill reindeer. In the Miocene there was a bear-sized wolverine that probably preyed on the abundant deer and antelopes of the plains.

One major line of the mustellids, the otters, became adapted to living in the water. The earliest otter, *Potamotherium*, comes from the early Miocene when it already looked and lived like its modern descendants. Between 1 and 1.5m long, its arched back was higher than its head and it moved on land in a series of short leaps. Like other water animals it had a poor sense of smell but its hearing and sight were very acute. *Potamotherium* was a high speed swimmer, superbly adapted to its water environment.

The true seals probably evolved from a similar ancestor. They have no external ears and cannot bend their back legs forwards so that on land they can only move by humping their backbone and pulling themselves along with their front flippers. In the water, however they are completely at home though they still return to land every year to breed.

The second group of seals, the furred or eared seals, include sea lions and walruses. Recently discovered fossils seem to show that they may be descended from the true dog line, not from the mustellids. If so, the adaptations both groups have made are the result of convergent evolution.

In North America there were several specialized dogs as well as the more familiar wolves, foxes and coyotes. *Daphoenus* was a bear dog, a long-tailed, omnivorous mammal that filled the niche of true bears (also members of the dog group). The osteobores were scavengers, like the hyaenas of the Old World. Otters such as *Potamotherium* and raccoons (*Procyon*) evolved from the primitive long-bodied mustellids.

Potamotherium

Procyon

Beavers, now well adapted to life in water, were once burrowing land animals. *Palaeocastor* lived in open grasslands during the Miocene and its spiral burrows are often preserved. Their complex pattern can be traced because the material with which the tunnels were eventually filled in is different from the surrounding soil.

The rodents

The rodents

Rodents were among the first mammals to evolve after the dinosaurs died out 64 million years ago. Small, inconspicuous animals with teeth and jaws specialized for gnawing and nibbling, they are now, in numbers at least, the most successful of all mammals.

Rodents are able to adapt quickly to changing environments and because they breed frequently and produce large litters, any successful characteristic can be rapidly passed on from one generation to the next and spread throughout the whole population. An elephant with a new characteristic cannot breed until it is at least 15 years old and will then pass its genes on to only one offspring. It will therefore take thousands – even millions of years – for the whole group to change. Rodents on the other hand are mature at 10 weeks, mate several times a year and produce a litter of at least 5 young. In only a relatively short time adaptations can be spread widely.

The basic rodent ancestor was the squirrel-like *Paramys* which had evolved in the Palaeocene of North America over 60 million years ago. By the Pliocene squirrels, true chipmunks, prairie dogs and marmots were flourishing on the North American prairies. Rodent ancestors must also have been alive in Eurasia for squirrels had evolved there too by the Pliocene. Another evolutionary line led in the Oligocene to land-living beavers and, later, to modern water-living types.

In South America and Africa rodents evolved in isolation: guinea pigs, capybaras and New World porcupines in South America; cane rats, hairless burrowing mole rats and Old World porcupines in Africa. New and Old World porcupines are not related – they have evolved in the same way, the New World living in trees, the Old World on the ground.

The most successful of all rodent groups are the rats and mice, which include voles, dormice, lemmings, jerboas, hamsters and gerbils and many more – as well as the familiar pests, the rats and mice themselves. Like the other groups, they originated from squirrel-like ancestors. Early voles were alive in the Oligocene in both North America and Europe and by the Miocene there were modern dormice and jerboas, hamsters and gerbils, insect-eating mole rats and pocket gophers. In the Pliocene literally hundreds of different vole species appeared, many of which survive today.

True rats and mice evolved in the Miocene. Some have been found in Europe but most seem to have been living in tropical Africa and Asia. It was not until the Pleistocene that they really began to increase, but today they are more numerous than all the other groups of rodents together. A large part of their success comes from their association with man. By taking advantage of his habit of storing food they have become one of the few groups to succeed not in spite of but because of man. They, like us, are opportunists.

The primates

The animals that developed most significantly during the Miocene were the advanced primates. During the early part of the Tertiary period, they had been confined to South America and Africa, where they had evolved in isolation. In Eurasia and North America the primitive tree shrews and tarsiers from which they had descended continued almost unchanged. In South America the New World monkeys remained forest dwellers but in Africa there were further changes.

The most important development of the early hominoids (gibbons, apes and man) was that their front limbs became much stronger, allowing them to swing through the trees by their arms alone. This is called brachiation. When an animal's weight is supported by its arms in this way it hangs upright with its legs in a straight line instead of at right angles to the body. This in turn leads to other changes. The position of the animal's head on its backbone has to change so that its face points forwards; and because it lands feet first, its feet and hands develop different functions. The feet simply support the weight of the body, while the hands, already used for grasping food, become more specialized for co-ordinating action with sight. Because their arms are long and strong, the animals are likely to become knuckle walkers on level ground, but they can also stand upright and walk on just their hind legs.

Many of the characteristics of modern primates began as adaptations for life in woodland and forest. The highly developed sense of balance needed for two-legged walking was essential for life in the trees and it was because a sense of smell was less important than sight and hearing that primates' faces became flatter.

As the grasslands spread and the forests thinned out, the normal habitat of the monkeys and apes was destroyed. At the same time, however, there was a new environment for them to exploit – the savannah. Some of the Old World monkeys which still used all four legs for climbing were able to adapt easily to life in the savannah; among the most successful were the baboons.

Notharctus

Squirrel monkey

Golden marmoset

Hanuman langur

Hamadryas baboon

Monkeys evolved in Africa and South America while both continents were isolated from the rest of the world and from each other. The early primate *Notharctus* is thought to be a common ancestor. In South America the New World monkeys are divided into two families, the squirrel-like marmosets and the cebids, which use their muscular tail as a fifth limb. Old World monkeys are all included in one family but include tree livers such as the langurs and larger ground-dwelling animals such as the baboons.

Living on the savannah

For survival, let alone success in an unfamiliar grassland habitat, individual animals had to learn to cooperate in organized communities. Baboons are a good example. When they are on the move, the females and their young travel with the dominant males in the centre of the group while other adult males and females without young form a protective guard. Juveniles travel at the outer edge of the group but not in the front or at the rear. When danger threatens the juveniles retreat to the centre and the dominant males move forward to confront the enemy While they stand to fight, the females and young make good their escape. Few predators can resist a group of angry, dominant males and the baboons usually succeed in driving off their attackers. Single animals are much more vulnerable and are often killed by the larger flesh eaters.

Baboons can themselves kill other animals such as small antelopes but they do not hunt co-operatively. Like other monkeys they are omnivores and will take any food, both animal and plant, whenever it is available.

The primitive apes, descended from the tree-living *Aegyptopithecus*, also took to life on the savannahs. They must have been among the most vulnerable of all the animals of the grasslands. They were not fast runners and they seem to have had no effective way of defending themselves. Yet in spite of this they became one of the most successful mammal groups in the new environment. When Africa was once more linked to Europe and Asia, these ancestral apes spread across into Europe, Asia, India and as far as China, proving that they were well able to hold their own against predators.

From their skulls we know that they had better brains than any other animals of their time and their success was probably largely due to their intelligence. They must also have been co-operative, social animals. One advantage they had over other mammals was that they could stand upright. This would have given them a good view of their surroundings so that they would have ample warning of possible danger and be able to run quickly out of the way.

These early apes belong to the genus *Dryopithecus*, which was first found in France in the last century. They must have been close to the common ancestor of both apes and man, but a detailed study of their jaws and teeth proves that they were already evolving along the line that lead to apes, not on the one that lead to man. They are probably the ancestors of modern chimpanzees and gorillas which finally evolved into their modern forms in the Pleistocene epoch.

The ape-like *Aegyptopithecus* was probably the common ancestor of both gibbons and apes. Its fossils first occur in Oligocene rocks of between 26 and 38 million years ago. The first gibbon, *Pliopithecus*, evolved in the Miocene. It already had the gibbon's characteristic long arms, an adaptation for moving through the trees. *Dryopithecus* and its close relative *Ramapithecus* also both evolved in the Miocene. They lived in more open country than the other primates and some may have begun to move freely on the ground. *Dryopithecus'* descendants were the true apes; *Ramapithecus* may be the first creature on the evolutionary line that leads direct to man.

Aegyptopithecus

The ancestors of man

A much rarer early ape from about the same period was named *Ramapithecus*. The first skull was discovered in India and described by a student, G. Lewis, in 1934. Lewis was certain that it belonged to a man-like ape but he was attacked vigorously by one of the leading anthropologists of the time and his ideas were quickly forgotten. It was not until 1964 that an American, Elwyn Simons, proved Lewis had been correct. Since then *Ramapithecus* fossils have been discovered in Africa, Turkey, Greece and Hungary.

Ramapithecus's jaws were more man-like than the other apes and its front teeth were smaller, hardly larger than its molars. It was a weaker and even more vulnerable animal than the more ape-like *Dryopithecus* and during the Miocene was much less common.

Ramapithecus probably lived on the fringe of the forest and savannah. Some chimpanzees live there today and their behaviour shows how the apes may have evolved. These savannah-living chimpanzees are partly meat eaters and, unlike the baboons, which only eat meat if they come across it accidentally, they hunt together

Pliopithecus

Ramapithecus

Dryopithecus

How the primates are related

actively. Moreover, they share their kills in a co-operative way. *Ramapithecus* may have behaved in the same kind of way but since it had no obvious weapons such as claws, and its teeth were smaller than a chimpanzee's, it must have found a new way of killing its prey.

One of the key characteristics of primates is their long thumb, which can be moved across towards the fingers to grasp objects firmly. They are also able to co-ordinate their hand movements with what they see and many are good at throwing. Even in the wild, chimpanzees use pieces of wood or stone to probe for termites in termite mounds and they have been known to make sponges from leaves to collect water they cannot reach. Chimpanzees only use tools like these when there is an immediate need to do so; unlike man, they do not make them for future use.

It is possible that *Ramapithecus* began to use sticks and stones not just as handy tools, but as actual weapons. They would have been dangerous extensions of the hand, much more effective than ape teeth and would have given *Ramapithecus* a definite advantage over the other early apes The late Louis Leakey found broken animal bones in Miocene rocks in Kenya which he thought had been damaged by the weapons of early apes. If he was correct, *Ramapithecus* marks the beginning of human stock not only in its physical features but also in its hunting behaviour.

There are two ways in which these man-apes could have obtained their meat. First, they could have acted like hyaenas, scavenging prey that had already been killed by one of the larger flesh eaters. Hyaenas can drive a leopard away from its prey as soon as it has made its kill and even a group of unarmed men could do the same.

The second, less dangerous method of hunting is to concentrate on killing small mammals. One of the main features of grassland is that it can support an enormous variety of animals. In the Miocene the many different types of hoofed mammals were the most obvious examples and many still remain on the plains of East Africa. However, the spread of the grasslands also brought a spectacular increase in the rodent group and these small mammals probably supplied most of *Rama-pithecus*'s animal protein needs.

85

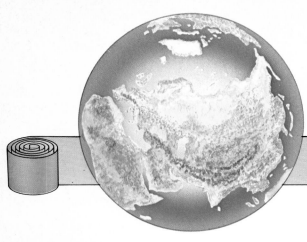

The Age of Man begins

The Age of Mammals began to go into a gradual, almost imperceptible decline as the Pliocene period progressed. The explosion of animal life that had taken place when the grasslands spread in the Miocene was over and as the climate cooled, the mammals became less varied.

Towards the end of the Pliocene, about 2 million years ago, there is abundant evidence of man's ancestors and by the middle of the Pleistocene, over a million years ago, he had established mastery over the rest of the animal world. As this one species with the ability to control all other living things became established, the Age of Mammals ended: the Age of Man began.

The pattern of animal evolution now began to change. In earlier times, evolution had produced a great variety of highly specialized mammals; now there were fewer but much more adaptable types. The result of both patterns of evolution is that a large number of ecological niches are fully occupied. In the old system, animals were specialized for a particular niche and could live in one environment. In the new system they were not specially adapted for any single way of life and so were able to survive in a wide variety of conditions.

Only on one continent did mammals continue to live in much the same way as they had for twenty million years. The plains of East and southern Africa still preserve an essentially Miocene pattern of mammal life.

This changeover in the pattern of animal life was a slow, gradual process that began with the man-like ape *Ramapithecus* and reached its peak in modern man, *Homo sapiens*. Other groups of animals followed a similar route. Dogs and cats, both highly intelligent and adaptable families, are able to cope with a wide range of situations. Among the plant eaters pigs, sheep, goats and cattle are the most successful. Finally, probably the greatest opportunists of all are the rats and mice.

The first men

The study of the remains of fossil man is a confusing one because in the very early stages there seem to have been several types of man-like creatures living at around the same time. Some scientists see these as completely separate species and consider that most died out, leaving one as the direct ancestor of man. But a simpler theory, based on the new pattern of evolution, is that they were all members of one very varied species, which included individuals of very different shapes and sizes – just as the human population does today.

Ramapithecus, the first really man-like ancestor, lived nearly 20 million years ago in Africa and Europe and by 7 million years ago had spread into Asia. During the Pliocene, it evolved into a slightly more advanced animal, *Australopithecus*.

The first *Australopithecus* skull was described in 1925 by Raymond Dart in South Africa. Its brain was only about the same size as a chimpanzee's but its jaws and teeth were more like a human's. Dart could tell from the place in the skull where the spinal cord had been that *Australopithecus* stood and walked upright.

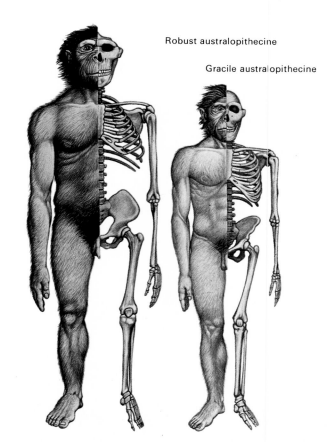

Robust australopithecine

Gracile australopithecine

Many more *Australopithecus* fossils have since been found and although scientists still disagree about many points, they all accept that there were two basic types – one lightly, the other more heavily built. They are known as gracile and robust australopithecines.

Australopithecines at different stages of evolution seem to have been living in the same areas at much the same time and some fossils have a mixture of primitive and advanced features. One of the oldest and most complete skeletons comes from Hadar in Ethiopia and is over 3 million years old. Known as 'Lucy', it has some features like *Ramapithecus*, some like *Australopithecus* and even some like a modern human's.

Although the number of different types is confusing, it is not really surprising. In a growing and evolving group, individual animals at different stages can be alive at the same time. No evolutionary change takes place suddenly or in every member of a particular group at the same time.

Physically, the australopithecines were not basically different from modern man. The only way to find out how 'human' they were is to work out how they lived.

At Makapansgat in South Africa some caves were probably a living site. Here there are many bones of small mammals with the front limbs of larger animals

The man-like australopithecines evolved during the Pliocene between 5 and 6 million years ago. Fossils of two distinct types have been found, one much more heavily built than the other. Several different theories have been put forward to account for the differences. They may have been males and females – or just very different types – of the same species; or two different species living at the same time; or even vegetarians and meat eaters. Whatever their relationship was, they probably lived as hunters on the grasslands and we know that at least some of them built permanent shelters and made primitive tools such as these stone choppers from Olduvai in East Africa. The hut (above) is based on a circle of stones, also from Olduvai. Similar huts are still built in the region today.

such as antelopes. Front legs are quite easy to remove even with a simple stone tool because the shoulder bone is attached to the backbone by muscle alone. Back legs are more difficult to cut off and only a few of these have been found at the site. This probably means that the australopithecines had to work quickly to save their kill from hyaenas and other scavengers.

Raymond Dart, who studied the site, suggested that the limb bones had been used as weapons and he found several baboon skulls which seem to prove this. They look as though they were broken by a long bone, held by a right-handed animal. Many of the bone fragments at Makapansgat look as if they were made for particular purposes such as cutting and scraping and there are also crudely made stone tools.

The site at Olduvai Gorge in East Africa where so many early human remains have been found also had simple pebble tools. Scientists do not agree about which type of australopithecine made them, but the main point is that they *were* manufactured. There is also evidence, the first ever discovered, of a man-made shelter: a ring of stones shows where a small windbreak or a round hut once stood.

It is from evidence like this that we are able to decide just how far the australopithecines had evolved. They were clearly hunters and scavengers; they lived in permanent 'homes' and they made tools for future use. They were therefore capable of thinking ahead and because of this they can be classed as primitive men.

The discovery of fire

The next stage in human evolution was first traced far away from Africa, in Java. In the second half of the nineteenth century, long before the australopithecines had been discovered, Ernest Haeckel decided that there must have been an animal part way between an ape and a man. A young Dutchman, Eugene Dubois, was so excited by this idea that he devoted his entire working life in Java to finding the so-called missing link.

Dubois was very successful. The fossils he found were about 750,000 years old, skulls with heavy brow ridges and low sloping foreheads and a leg bone that was exactly like that of a modern man. Later, in 1927, more fossils were found in Peking. These were between 500,000 and 250,000 years old but were eventually included in the same species, *Homo erectus*.

The fossils showed that *Homo erectus* people varied considerably and that not all had the low, sloping forehead of the first skulls that had been found. Their average height was about 1.5m and they were rather stockily built. They had no chin and probably could not speak very well for the muscles controlling the voice box or larynx were short, like those of a baby or toddler today. Nevertheless they could probably make a variety of sounds and the beginning of communication by language could have begun at this stage.

Homo erectus was a much more successful type of early man than the australopithecines. He still made bone and stone tools but instead of using simple chipped pebbles, he made larger hand-axes, the basic all purpose tools of early man. His most important step, which was to have a fundamental effect on the rest of the animal world, was the discovery and control of fire.

Fire is one of the few natural phenomena that panics wild animals and if controlled can be used to drive them in whatever direction the hunter wishes. Many smaller animals would have been accidentally burned during a fire and the advantages of this would soon have been seen. They would not only have been more pleasant and easier to eat, but in a cooked or dried state would keep much longer than freshly killed meat.

Early man in Europe

The great Ice Ages, which began over 2 million years ago, seem to have stimulated man's evolution in the northern lands. Because he could survive in colder climates he was able to move further north, even as the general climate there grew harsher.

In 1960-1 Clark Howell excavated a site at Ambrona about 160km north-east of Madrid in Spain. Here he found direct evidence of how *Homo erectus* hunted.

The site was a narrow valley, which must have been a regular migration route for elephants. We can tell from the sediments that the area suffered severe frosts and the elephants would have been travelling south to warmer regions. Bands of hunters apparently waited for the herds of elephants to appear and then lit fires to drive them into boggy ground. Fragments of charcoal and carbon show that the fires spread over wide areas. The elephants, trapped in the soft marsh, were killed with rocks and spears and cut up on the spot. Meat from the carcases was divided up among the hunters and cooked or dried on small fires before the groups dispersed. Beside the remains of the fires are piles of broken bones which include other mammals as well as elephants. The important thing about them is that the different heaps all contain the same types of animal bones. Obviously, the animals they caught were shared out equally among the hunters. This is common practice among modern hunting people but it is interesting to find that it began at this early stage of human evolution – over 250,000 years ago.

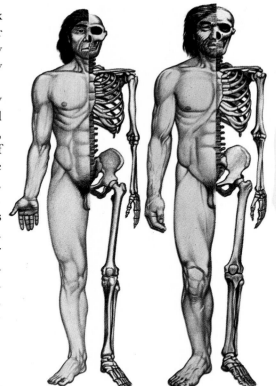

Homo erectus   Neanderthal man   Cro-Magnon man

Homo erectus was able to deal with even the largest prey animals and he probably also travelled to follow seasonal migrations. A summer hunting camp was discovered in 1966 in Nice, on the south coast of France. This site, known as Terra Amata, seems to have been occupied from time to time over hundreds of years.

By about 200,000 years ago, early man had become *Homo sapiens*, modern man. The main physical change seems to have been an increase in brain size, with a higher and more rounded forehead, though still with heavy brow ridges. By far the most significant advance was shown in the kind of tools that these people could produce. The workmanship was much more delicate than before, implying that they had learned to use a precision grip. With a precision grip objects can be held between the thumb and the index finger, as a pen is held for writing. *Homo erectus* could not do this; he had only a power grip, where an object is held firmly between the thumb and the rest of the fingers, and so could make only relatively rough tools.

The early *Homo sapiens* people are known as Neanderthals, the name of the valley where their first remains were discovered. They were well adapted to life in a cold climate. They built small skin-covered tents inside caves and were better hunters than their ancestors.

Neanderthals also seem to have had some kind of

Neanderthals' flattened face and heavy build were probably a protection against the bitter cold of the Ice Ages. With wooden lances and stones lashed to wooden handles they were able to hunt the large mammals that roamed the European tundra. The meat provided them with food; the skins were used as clothing.

Above left: The first *Homo erectus* fossils are 1.75 million years old, the last less than 250,000. Our own species, *Homo sapiens*, seems to have evolved nearly 200,000 years ago. Neanderthal man was living in Europe 100,000 years ago, during the great Ice Ages. He was succeeded about 40,000 years ago by Cro-Magnon man – truly modern people.

religious sense for they buried their own dead in shallow graves in the cave floors. One of the most striking recent discoveries is that before the body was covered, bunches of flowers were placed on it.

Although they are included in our own species *Homo sapiens*, Neanderthals still had heavy brow ridges like *Homo erectus* and had not yet developed the modern type of chin. One group, the Classic Neanderthals, were stocky, highly adapted to harsh life in cold conditions – rather like today's eskimoes. They were replaced in western Europe by another group, the Cro-Magnon people. These were more varied than the Neanderthals, had smaller brow ridges and developed a modern, jutting chin. Although Neanderthals disappeared, many of their cultural traditions seem to have continued.

It is often said that the Neanderthals became extinct, but this is unlikely to be what actually happened. Some living human groups still have physical features which in fossils are a sure sign of Neanderthal man. The Neanderthal's occipital 'bun', a rounded, bony lump on the underside of the back of the skull, is quite common today – the author of this book has one! This shows that although Neanderthals disappeared as a separate group, they must have interbred with other types such as Cro-Magnon people, for their genes still survive in modern man.

Gibbons, apes and man

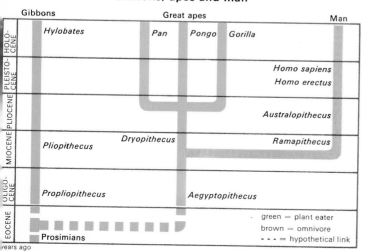

	Gibbons		Great apes			Man
HOLOCENE	Hylobates		Pan	Pongo	Gorilla	
PLEISTOCENE						Homo sapiens
						Homo erectus
PLIOCENE						Australopithecus
MIOCENE	Pliopithecus		Dryopithecus			Ramapithecus
OLIGOCENE	Propliopithecus		Aegyptopithecus			
EOCENE	Prosimians					

green = plant eater
brown = omnivore
- - - = hypothetical link

years ago

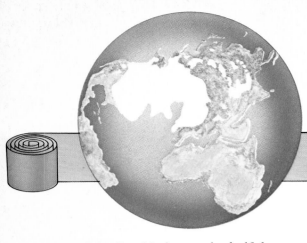

The Pleistocene Ice Ages

For most of the Earth's four and a half thousand million year history there have been no polar ice caps. During the last two million years, however, there have been seven Ice Ages, interrupted by shorter warm periods, the interglacials. During this time the temperature changes from warm to cold and back to warm again were more extreme than at any other known period. Today about 10% of the Earth's surface is covered by ice, making up a volume of 26 million cubic kilometres and locking up 75% of the world's freshwater. Towards the end of the last Ice Age, 15,000 years ago, ice covered 30% of the surface, with an estimated volume of 76 million cubic kilometres.

The world climate had been cooling gradually since the Tertiary and during the Oligocene epoch a relatively small ice cap had developed at the South Pole. During the Pleistocene the cooling process suddenly speeded up. Vast polar ice sheets spread towards the equator and the cold zones of permafrost and tundra reached about 2000km further south than they do today. Deciduous woodlands also grew nearer to the equator while in the equatorial regions themselves dry desert areas formed.

As millions of cubic kilometres of water were locked in the ice caps, the sea level fell by about 90m. Large areas of the continental shelves became dry land where both plants and animals could live. Old river beds, now hundreds of kilometres out at sea, show how far the land extended. When the ice caps melted during the warmer interglacials, the new land was flooded again. This continual succession of hot and cold climates affected plant distribution and must also have had a strong influence on animal life.

The tundra is the land immediately south of the polar ice cap. Covered by ice and snow in winter, much of the underlying soil is frozen all the year round. Towards its southern edge are patches of pine forest and willow scrub while at the northern boundary the bare rocks are covered only by lichens. In between are areas of marsh, moss, moorland vegetation and sparse grassland.

In the Ice Ages the northern tundra stretched as far south as the Mediterranean in Europe, the Himalayas in Asia and the southern States of North America. Animals adapted to cold climates increased while the ones that could only live in more temperate areas were forced to migrate southwards.

On the tundra there are periods of the year when food is very scarce indeed, but others when there is plenty to eat. Animals adapt to this situation in two ways. Some, like caribou and reindeer, migrate to an area where there is enough food in the winter. Others, like the Ice Age cave bears, spend the winter in hibernation, surviving on the food stored in their body during the summer months.

Whether they sleep or migrate for the winter, tundra-living animals must be opportunists, able to use any

Woolly mammoth

Musk ox

Woolly rhinoceros

Many of the large plant eaters became well adapted to the extreme cold of the Ice Ages and grew thick hairy coats. The woolly mammoth and woolly rhinoceros are the best known; their exact appearance was recorded by early man in his cave paintings. The musk-ox, a relative of the sheep, still survives in the Arctic today.

food that is available. The environment itself is less varied than temperate grassland and provides fewer different habitats. No animal with highly specialized food requirements would stand much chance of survival.

Surprisingly, as the tundra spread, the actual weight of animals for every square kilometre of land increased in much the same way as it had when the grasslands spread in the Miocene. But whereas the Miocene changes had brought an explosion of new species, in the Pleistocene Ice Ages adaptability and variation within a few species became the most successful pattern.

As the ice caps advanced and retreated, several mammals used to more temperate climates changed so that they could survive in colder conditions. Many became fully adapted to life on the tundra. The most familiar of these is the woolly mammoth, an elephant with a high domed head and back, thick, shaggy hair and open spiral curved tusks. These great tusks could have been used to sweep away soft snow and uncover the plants.

There must have been enormous numbers of these woolly mammoths and their fossil tusks were a major source of ivory in mediaeval times. Their anatomy is very well known for complete animals have been preserved deep frozen in the permafrost, with even parts of their last meals in their mouth. Even details of the hair of their thick coats are known.

Until the last Ice Age, man was very much a part of the rest of the animal world. As a nomadic hunter he followed the large game animals in their annual migrations to and from the feeding grounds. When the woolly mammoth evolved, however, it provided an abundant source of food which remained in the same general region. During the last Ice Age settled hunting communities depending entirely on this one species developed.

Another mammal now found only in warmer countries, the rhinoceros, developed a woolly form adapted to Arctic conditions. Living sub-arctic mammals such as musk-oxen, reindeer, hares, lemmings, wolves and bears were all common in the Ice Ages and the ancestor of modern cattle, the aurochs, had evolved. In North America another elephant, the woodland-living mastodon, evolved a thick, hairy covering as an adaptation to the extreme cold.

Many fossils from the Ice Ages are preserved in caves, where they were protected from the scouring effect of the ice that covered so much of the northern lands. Some of the caves had been used by hyaenas as dens and they contain the remains of prey animals and of small scavengers.

Caves inhabited by cave bears often contain their bones as many died in their sleep during their annual

During the last Ice Age many of the caves in Europe were inhabited by large cave bears. They survived the bitter winters by hibernating but many died in their sleep and the species became extinct before the Ice Age ended. Cave bears seem to have had a special significance for Neanderthal man; he competed with them for the shelter of the caves and often buried their heads in special graves in the cave floor.

hibernation. One cave in Austria contains the remains of 30,000 bears. Some of the older bears suffered badly from osteoarthritis. This would have crippled them, making it difficult for them to find enough food in the summer to build up the layer of stored food they needed to see them through the winter.

Sometimes the roofs of these caves collapsed, forming deep natural shafts. Animals often fell down, to be fossilized among the rubble on the cave floors. Scavengers that managed to clamber down to feed on the carcases found it impossible to get out again and they, too, died and were eventually fossilized.

The Ice Age giants

Perhaps the most remarkable development of the mammals during the Ice Ages was the evolution of giant forms, which appeared in both the cold and the tropical areas. In North America the beaver *Casteroides* was 2.75m long, nearly three times the length of a beaver today; a giant warthog *Afrochoerus* from Africa was 3m long, twice the size of the living warthog and the Asian grazing rhinoceros *Elasmotherium*, with its 2m long horn, was almost twice the length of a living rhinoceros. Also in Asia, the largest ever primate, the gorilla-like *Gigantopithecus*, was nearly 3m tall. Horses, bison, buffalo, aurochs, deer (in the form of the giant Irish elk *Megaceros*), cheetahs, wolves, mammoths and elephants were

all larger during the Ice Ages than at any other time.

Giant forms of modern animals were certainly one of the most striking features of the Ice Ages. However, when a particular species or genus is traced from one cold period to the next via the warm interglacial we find that its overall size varied with the climate.

For some animals, including the ungulates and many of the flesh eaters, giant size was an adaptation to reduce heat loss. A larger bulk means an animal has a smaller relative surface area and will take longer to lose heat – especially if it is well insulated with thick fur.

Smaller animals such as ermines and stoats followed an opposite pattern: they were smaller in the cold periods and larger during the warm interglacials. This was probably because more food was available to them in warmer times. Surprisingly, the woolly mammoths also became larger during the warmer periods when there were more suitable plants for them to eat, and smaller in the cold. In the warmer areas of the world this pattern seems to have been the most common one.

Mammals in the interglacials

During the warm interglacials, mammals from the warmer areas moved northwards, replacing the woolly mammoths, and rhinoceroses, musk oxen and reindeer which in their turn moved still further north. In each interglacial, the animals were very similar.

England was obviously a much warmer place than it is today, with average temperatures more like those of southern Europe. A fascinating selection of animals was discovered during building excavations in Trafalgar Square in the centre of London. The sediments in which the fossils were found had been laid down by the River Thames, which was very much wider than it is today. Remains of plants and insects were found as well as fossils of mammals.

The plants would not look out of place there today but the animals would be quite a surprise. They included straight-tusked elephants, rhinoceroses and hippopotamuses as well as red and fallow deer, aurochs, bison and wild pigs. Among the carnivores were cave lions, brown bears and hyaenas. The people who were living there at the time would have had no shortage of prey animals, but they would have had to be always on the watch for their fellow predators.

Some 40,000 years ago man and a few other mammals crossed the Bering land bridge from Asia to North America. At this time the ice caps were not large enough to make the far north uninhabitable but they had locked up enough water to lower sea levels and expose the land. Man also reached South America and Australia by land but as the ice retreated and the sea rose again, the early migrants were cut off and the continents isolated once again.

Next page: Trafalgar Square, London, during the last interglacial about 100,000 years ago. The straight-tusked elephant *Palaeoloxodon* (1) lived alongside wild pigs aurochs (2) bison, red and fallow deer (3). Hippopotamuses (4) wallowed in the shallow waters of the Thames. The flesh eaters included cave lions (5), brown bears and hyaenas. Even fossils of the small rose chafer beetle (6) have been found.

The islands: giants and pygmies

As the Ice Ages and interglacials of the Pleistocene came and went, sea levels rose and fell all over the world. When the ice caps expanded more land in the north became uninhabitable, but around the coasts new land appeared and was quickly colonized by the mammals. When the ice melted, the new land was flooded, leaving animals isolated on islands of higher ground. The mammals on these islands belonged to the same groups as those on the mainland but instead of producing giant versions they evolved in the opposite direction and became pygmies.

One of the main reasons for this is that there were no flesh eaters on the islands during the Pleistocene. On a small island with no predators to control the numbers of plant eaters there is only a limited supply of food. Large animals such as elephants would quickly destroy the habitat and, with nowhere to migrate, would die out. Only the smaller ones, needing less food, would have a chance of surviving.

The islands off the coast of Indonesia and in the Celebes were inhabited by dwarf stegodont elephants and in Santa Barbara off the coast of California there was a race of small woolly mammoths. The smallest elephant of all, a miniature version of the straight-tusked *Palaeoloxodon*, lived in Malta and was 1m tall.

The other group of miniature mammals common to many islands were pygmy hippopotamuses – about half the size of today's animals. In Malta there was also the 'giant' deer *Megaceros*, which there stood only 1.5m. On the Antilles in the Caribbean, the South American ground sloth (usually some 4m high) ended up little larger than a tabby cat.

While the larger mammals were growing smaller, some of the smallest ones were actually becoming giants. Giant pikas evolved in Sardinia and Corsica and on other Mediterranean islands there were giant dormice and rats. These rodents were probably able to grow larger because there were no predators to avoid and it was no longer an advantage to be small and inconspicuous. They were much larger than rats and dormice on the mainlands but still only about a quarter the size of the islands' small elephants. Clearly there must have been enough plant food to support them.

Madagascar

Unlike the small Pleistocene islands, Madagascar has been separated from the east coast of Africa by the Indian ocean for millions of years, since the very beginning of the Age of Mammals. The only mammals to reach it were primitive tree-living insect eaters and

Palaeoloxodon

Megaceros

Megatherium

When sea levels rose during the interglacials many animals were isolated on new islands. On Mediterranean, Caribbean and Indian Ocean islands some mammals evolved miniature versions which were able to survive on the limited amount of food available. There were dwarf elephants (*Palaeoloxodon*), minute Irish elks (*Megaceros*) and cat-sized giant ground sloths (*Megatherium*). In contrast, some normally small animals grew into giants – Maltese dormice and rats were around 25cm tall.

primates. They probably arrived there by accident, clinging to floating trees washed out to sea. Later, pygmy hippopotamuses also crossed from Africa.

Free from competition and, at first, from flesh eaters, the early primates (the lemurs) and the insect eaters (the tenrecs) evolved into varied groups of animals that are still found only in Madagascar.

The lemurs are all tree living but avoid competition by feeding at different times and on different types of food. Some forage in troops at early morning and late evening; tiny mouse lemurs are nocturnal; dwarf lemurs sleep through the dry season, when food is scarce, living off fat stored in their tail. The most dramatic looking of the lemurs are the short tailed indris and sifakas which are more monkey-like, about 1m long. During the Pleistocene there was a giant lemur, *Megaladapis*, about the size of a chimpanzee. *Megaladapis* survived until the eighteenth century.

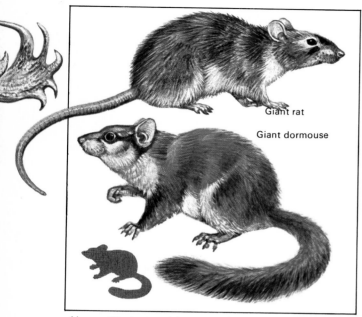

Above: Giant rat and dormouse drawn to the same scale as a mainland form (in silhouette).
Below: Mainland elephant drawn in scale with island pygmies (elephant, elk, ground sloth) and giants (dormouse and rat).

Mammals of Madagascar

Aye-aye

Tenrec

Mouse lemur

The island was isolated during the early part of the Age of Mammals. Primitive primates, the lemurs (above) and insect-eating tenrecs were the main mammals. Both groups evolved many different forms.

Until the arrival of man, civets such as the fossa and mongooses were the only flesh eaters on the island. Together with the voles, they managed to reach Madagascar during the Miocene.

Fossa

Mongoose

Vole

The most highly specialized of all the lemurs is the aye-aye. It fills the ecological niche of the woodpeckers which do not live in Madagascar. The other lemurs have long lower incisor teeth which they use like a comb for grooming their fur. The aye-aye has chisel-like 'rodent' incisors. It also has very large ears and a long, bony third finger. Clinging to the trees with its back legs, it taps the bark with this long finger and then listens to see if insects or grubs are inside. If they are, it gnaws away the bark or wood with its teeth.

Lemurs are closely related to the bushbabies of Africa and the small nocturnal tarsiers of Asia. Like them they are closer to primate ancestors than to more advanced monkeys and apes.

The insect-eating tenrecs also evolved to fill different ecological niches, this time on the ground. The basic tenrec is a small-brained animal, an early descendant of the first shrew-like mammals that evolved over 200

million years ago. Now there are some thirty separate species, ranging from a spiny hedgehog-like form to long-tailed species like shrews, a water-living type that lives on shrimps and one that burrows like a mole.

During the Miocene two further groups of mammals managed to reach Madagascar: primitive members of the cat group, the viverrids and an early type of vole. Today the viverrids include three species of civet (including one, the fossa, the size of a fox) and nine types of mongoose. These are still the only true carnivores in Madagascar. The voles, which arrived around the same time have now evolved into fifteen species.

Before man arrived in Madagascar it had remained almost unchanged for over 60 million years. Its tropical forests were very like those of North America and Europe during the Palaeocene, before the spread of the grasslands and its primitive primates and insect eaters allow us a glimpse into an earlier world.

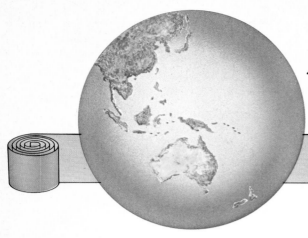

Australia: the Island Continent

From about 64 million years ago until the middle of the last Ice Age, Australia was quite cut off from the rest of the continents. Like a large, isolated island, it provided an environment where animal life – the mammals in particular – evolved in a unique way.

The most primitive mammals of all, the egg-laying monotremes, are found today only in Australasia, where they have lived almost unchanged for around 200 million years. Monotremes are classed as mammals mainly because they feed their young on milk produced by the mother. In other ways they are more like reptiles: they lay shelled eggs and have small brains. They may in fact be surviving members of the ancient paramammal group which died out elsewhere around 200 million years ago.

Possible monotreme ancestors have been found in 200 million year old rocks in South Wales but there is no further hint of their existence until around 5 million years ago when fossil echidnas are known in Australia. During the Pleistocene there were giant echidnas and large duck-billed platypuses; both these groups still survive today. Though they are primitive mammals, they have established themselves in secure ecological niches and there is little chance that they will become endangered species.

Australia's best known animals, however, are all marsupials. At the time Australia broke away from the other continents placentals had not yet evolved and the marsupials were able to develop free from competition. Today Australasian marsupials live in Australia, Tasmania, New Guinea, Timor and Celebes.

The first known marsupials come from the late Cretaceous in North America, over 64 million years ago. By the Palaeocene they had reached both South America and Europe. The only way they could have reached Australia is from South America via the continent of Antarctica at some time early in their history, while these three land masses were still linked. Unfortunately no fossil mammals are known from Antarctica from either the Cretaceous or the Tertiary periods.

Marsupials died out in Europe over 7 million years ago and apparently never reached Africa. Several survived in South America until the Pleistocene, 2 million years ago but today the only groups outside Australia are the insect-eating rat opossums of South America and the scavenging opossums which have spread from the south far into North America. The rest have been replaced by placental mammals which are, on the whole, more intelligent and adaptable animals.

Modern Australian mammals are obviously the result of millions of years of evolution and change, yet the fossil record of their ancestors is very poor. Until the 1960s the only fossils were of Pleistocene ancestors – many, like mammals on other continents, growing to giant size. Now, however, early animals from the major marsupial groups have been found in rocks of either late Oligocene (30 million years ago) or early Miocene (26 million years ago). The evolutionary stages that lead up to these are still unknown.

Tiger cat

Marsupial mouse

Duck-billed platypus

Echidna

The flesh eaters and insect eaters

The most primitive marsupials were all flesh eaters and include the living opossums and the extinct borhyaenids of South America. This group is not found in Australia even though many Australian marsupials are popularly known as possums.

The earliest flesh eaters in Australia are the dasyurids, marsupial equivalents of cats and mice, which are first found in rocks around 30 million years old. They are tree living forest hunters. The Tasmanian devil *Sarcophiles* is a fierce, cat-like hunter and scavenger and the smaller dasyures feed on small mammals and birds in the tree tops. Both fill the ecological niche of placental civets in other lands and seem to have first appeared around the same time. The marsupial 'mice' are the smallest of all marsupials. They are not at all mouse-like in their habits but eat insects, filling the ecological niche of shrews.

Two other insect eaters, the marsupial mole and the numbat or banded anteater, are not found in the fossil record at all. The mole lives in exactly the same way as a placental mole, digging tunnels and burrows underground and feeding on insects and worms. The anteater is adapted to feeding on ants and termites and has a long, sticky tongue and powerful claws like placental

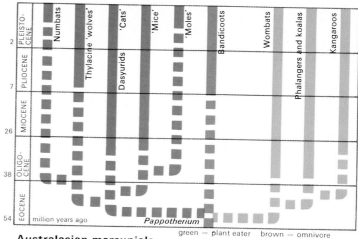

Australasian marsupials

green = plant eater   brown = omnivore
red = flesh eater   - - - = hypothetical link

Egg-laying monotremes (the spiny anteaters and duck-billed platypus) are the most ancient of all the surviving mammals. In many ways they are still like the ancestral paramammals which died out some 200 million years ago.

Marsupials are also primitive mammals compared with placentals but because Australasia has been isolated for so long, they have taken over all the ecological niches that are filled elsewhere by the more intelligent group. Marsupial mice, like placental shrews and hedgehogs, eat insects. Tree-living dasyures are the equivalent of civets and marsupial wolves such as the nearly extinct Tasmanian wolf live in much the same kind of way as true wolves. There is even a marsupial mole and a termite eater – the numbat. Bandicoots, like placental elephant shrews in Africa, are omnivores. Among the plant eaters, wombats are burrowers, like the extinct land beavers of North America; koalas are equivalent to small pandas and phalangers parallel squirrels and flying lemurs.

Many of the marsupials even look like their placental equivalents but the kangaroos and wallabies, which fill the niche of odd- and even-toed ungulates, are unique in every way.

anteaters. However, it is smaller and less extremely adapted than the anteaters from Africa and South America, looking more like a bushy-tailed squirrel.

The Tasmanian wolf *Thylacinus* first appeared during the Pliocene (7–2 million years ago). The extinct borhyaenids of South America must have been very like *Thylacinus* and may, in the far distant past, have evolved from similar ancestors.

Tasmanian wolves are the largest living marsupial flesh eaters. They are now about the size of dogs but in the Pleistocene they – and the Tasmanian devil – were probably up to 50% larger. They are specialized for running down their prey and like dogs or wolves they have great powers of endurance. Even though they may not be particularly swift, they can keep going until their prey finally tires and falls.

The larger marsupial 'cats' and 'wolves' usually prey on wallabies and kangaroos but man is rapidly destroying their habitats. The Tasmanian wolf is very rare today and may soon become extinct.

Bandicoots, another marsupial group, are also first known from the Pliocene. They look like African elephant shrews, with very long snouts and long, spindly back legs. Bandicoots may be related to the flesh eaters but their back feet are similar to a kangaroo's and they may be connected in some way with the plant eaters. Today they are omnivores, eating roots as well as lizards, worms and a variety of different insects.

The plant eaters

Australian marsupial plant eaters all belong to the same group, the diprotodonts, meaning literally 'two front teeth'. They are divided into three families, the phalangers, the wombats and the wallabies and kangaroos.

Australia's oldest fossil marsupial, *Wynyardia*, was a diprotodont plant eater, found in Oligocene sediments in Tasmania. By the Miocene, ancestors of all three families were alive and had begun to specialize for their different ways of life.

The first known phalanger was an early koala bear. Today koalas feed exclusively on *Eucalyptus* leaves but it is impossible to know whether their Miocene forerunners were as specialized as this. The other phalangers include a wide variety of leaf and fruit eaters, all living in the trees. Some have developed like the smaller tree-living primates: they use their tails as an extra hand like monkeys in the New World and the cuscus eats small mammals and birds as well as plant food. Other phalangers such as the sugar glider are more like rodent flying squirrels or the flying lemurs of Asia, and fill the same ecological niche as these placental mammals.

True wombats first appeared in the Pliocene and were already specialized for digging and burrowing. They are now about 1m long but in the Pleistocene there was a giant, *Diprotodon*, which was about the size of a rhinoceros, some 3m long. *Diprotodon* was much too big to be a burrower like modern wombats and probably fed on plants that were easily dug out of the soil.

Several skeletons of *Diprotodon* have been found preserved in lake muds. During the Pleistocene there were salt lakes in Australia, rather like the Dead Sea today. As the water evaporated in the hot sun, a thick crust of salt formed on top, strong enough to bear an animal's weight. Many *Diprotodons* must have been just too heavy for the crust, for it gave way as they walked across it and they fell into the water below. Their skeletons eventually sank into the muddy floor and were preserved whole.

The first kangaroos and wallabies come from Oligocene-Miocene rocks. These are highly specialized plant eaters with teeth adapted for cropping and chewing grasses. Some species have become tree-livers, but most live on the ground on open plains or woodland.

Kangaroos fill the same ecological niches as placental ungulates such as horses, antelopes, sheep and cattle do

During the Pleistocene there were giant marsupials in Australia, just as in other parts of the world there were giant placental mammals. *Diprotodon*, a wombat, was nearly 4m long, the size of a rhinoceros. *Procoptodon*, a short-tailed kangaroo, was over 3m tall. The marsupial lion *Thylacoleo* is now extinct. It may have been a flesh eater or scavenger but some scientists believe it fed only on fruit.

elsewhere. Like them they need to be able to move quickly to avoid predators. In other cases where animals lived in the same way they evolved similar shapes and ways of moving. South American plant eaters, for example, developed horse-like and camel-like forms quite separately from the north. Kangaroos and wallabies look very different from their placental equivalents and move in a series of leaps, propelled by their strong back legs.

During the Pleistocene there were giant forms of both kangaroos and wallabies and a short-tailed, short-faced form evolved which was probably a browser. This is now extinct.

Another extinct member of the diprotodont group was the marsupial 'lion' *Thylacaleo*. *Thylacaleo* may have been a flesh eater for it had strong stabbing and shearing teeth and a lion-like body, apparently well adapted for hunting. But its unusual teeth could also have been used for eating fruit, and it may, after all, have been a plant eater.

During the Pleistocene the first placental mammals reached Australia, presumably by crossing from island to island or by clinging to floating driftwood. The first immigrants were members of the rat and mouse family. Today there is a water rat which burrows in river banks and huts shellfish, frogs and fish; the stick rat which builds nests of sticks; and the small, delicate mouse *Leggadina* which is simply a tiny mouse, quite unlike the ferocious insect-eating marsupial mice.

Towards the end of the Pleistocene, some 30,000 to 40,000 years ago the sea was sufficiently low for man to cross from Asia. With him came the dog *Canis dingo*. Sea levels rose again as the Ice Age ended and few other placental mammals – except those introduced by man – managed to make the crossing. In spite of the destruction of their habitat by modern man, many of Australia's marsupial mammals seem to be holding their own.

Procoptodon

Diprotodon

Thylacoleo

The New World: Migrations and Extinctions

During the Ice Ages much of North America was, like Europe and Asia, covered by a vast ice cap which advanced and retreated as cold and warmer periods followed one another. Cold-loving mammals flourished around the edges of the ice and during the interglacials more temperate animals moved northwards.

Among the mammals adapted to the cold were the giant moose *Cervalces* and the musk-ox *Preptoceras*, as well as the American mastodon with its hairy coat. Bison were by far the most common plant eaters, living all over the American plains from near the edges of the ice cap almost to the Gulf coast. During the Pleistocene they drove most of the other North American ruminants into extinction. In the warmer periods horses, camels, pronghorns and giant peccaries also moved further north. The main flesh eaters were wolves and large biting cats such as the Californian lion. Sabre-tooth cats preyed on the large, slow-moving mammals such as the imperial mammoths and the dire wolves acted as scavengers – like the hyaenas of the Old World.

In warmer South America many of the unique mammals grew to giant forms, just as they did in other parts of the world. Three of the most famous mammals from this time were discovered by Charles Darwin in the 1830s. The skull of one was being used as a target for their stones by some young farm boys. Darwin called it *Toxodon*.

Toxodon was a heavily built animal, rather like a modern rhinoceros but without a horn. It was descended from the group now known as toxodonts which had evolved over 30 million years earlier.

Another of Darwin's discoveries was the long-necked litoptern *Macrauchenia*, which was like a camel but probably had a long trunk for browsing on leaves. *Macrauchenia*'s ancestors had been more like llamas but were replaced by true llamas from the north during the Pliocene.

Darwin's third major discovery was the giant ground sloths. During the Pleistocene these were real giants some, such as *Megatherium*, up to 6m high sitting on their haunches to feed on the tops of trees.

The final dramatic-looking South American mammals to evolve in the Pleistocene were the heavily armoured glyptodonts *Glyptodon* and *Doedicurus* – giant armadillos with thick curved shells and dangerous, spiky tails.

All these impressive mammals were flourishing when the most dramatic event in the history of South America took place. The land link with North America which had been broken at the beginning of the Tertiary period 64 million years ago, was fully restored for the first time. A narrow link had been formed in the Pliocene but thick tropical forests had covered the land and only a few forest-living mammals had managed to cross it. As the temperature cooled and sea levels fell, the isthmus of Panama formed a permanent land bridge between the two continents, allowing free migration in both directions from north to south and from south to north.

During the Ice Ages the North American plains were dominated by bison. These members of the cattle group had migrated across the Bering land bridge from Asia and soon replaced the many species of North American pronghorn. In the northern regions the giant moose managed to flourish throughout the cold periods.

The invaders

Throughout the Tertiary, for nearly 60 million years, the only flesh eaters in South America had been the different species of marsupials, the borhyaenas. Now the placental carnivores from the north invaded. Packs of wolves, large biting cats and sabre-tooths all moved southwards. The South American plant eaters were quite unprepared for their efficient, intelligent methods of hunting and many soon became extinct. The marsupial borhyaenas were unable to compete with the more intelligent and efficient placental hunters and as there were fewer plant eaters for the flesh eaters to share, they too, died out.

Plant eaters from the north also moved into South America, particularly one-toed horses and deer. Gradually they replaced many of the unique South American plant eaters. The largest North American animal, the imperial mammoth, lived successfully even in the high Andes. They survived longer than any of the other elephant species (except for the two living today) and an elephant hunt is known to have taken place in South America as late as 600 A.D.

Some of the invaders that were most successful during the Pleistocene (the elephants, sabre-tooth cats and horses) later became extinct but tapirs and spectacled bears still survive in their adopted home though they are no longer found in the north.

For some reason bisons, the most numerous of all North American ungulates, never migrated to the southern continent. Bison always travelled in large herds and it is possible that there were not enough grasslands to provide food for them on the journey. Their thick, shaggy coats, an adaptation to a cooler climate, would also have discouraged them from moving into warm, sub-tropical environments.

As the northern animals took over the southern lands,

Giant moose

Bison

South to temperate North

Megatherium

Opossum

Armadillo

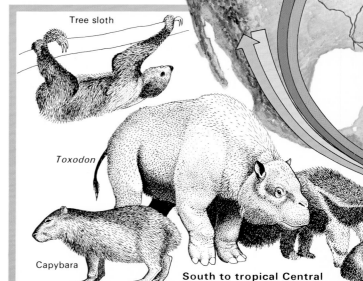

Tree sloth

Toxodon

Capybara

South to tropical Central

the unique groups of mammals that had evolved over 60 million years were doomed to extinction. But this was not the whole story. Many South American mammals managed to migrate northwards and became firmly established in North America. The most successful were the opossums which had in fact evolved originally in the north but had become extinct there. Today they are still extending their range, moving north into regions where no-one imagined they would survive.

The second successful south-north invaders were the porcupines which have established themselves in the north. Armadillos, too, reached the temperate southern areas of North America.

Toxodon, giant anteaters, capybaras, agoutis and spider monkeys also moved north but only as far as Central America where the climate was still tropical. *Toxodon* is now extinct and giant anteaters only survive in South America. The other immigrants live on in both their original and adopted lands.

Unexpectedly giant ground sloths and glyptodonts lived very successfully in the southern parts of North America for a considerable time. At first sight they seem particularly defenceless against fierce flesh eaters such as the Californian lion, the heavily built sabre-tooth *Smilodon* and packs of wolves and dire wolves. Yet they survived in the same territory as the carnivores for hundreds of thousands of years.

The tar pits

Towards the end of the Pleistocene, in what is now the centre of Los Angeles, oil seeping up from underground formed large, sticky pools of tar on the surface. Now hardened into rocks, the Rancho la Brea tar pits contain an enormous number of mammal bones, the skeletons of animals that became trapped and later buried in the sticky tar. They provide an interesting record of many animals of the late Pleistocene, up to 11,000 years ago.

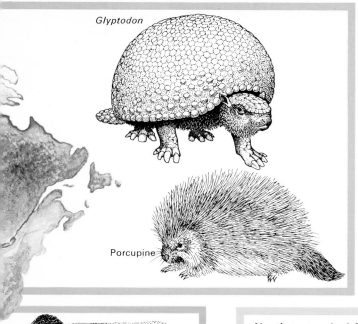

Glyptodon

Porcupine

The great migration

When South and North America were rejoined in the Pleistocene, animals migrated in both directions, seeking new territory. The invasion was disastrous for the native South American mammals. The northern plant eaters were generally more efficient than their southern equivalents while the marsupial flesh eaters could not compete with the more intelligent placentals.

Most of the northern invaders took over more temperate areas of South America and many, including spectacled bears, tapirs, wolves, pampas deer and hares still live there today. Horses and sabre-tooth cats died out in both north and south at the end of the Pleistocene but Imperial mammoths were alive in South America as recently as 600 A.D.

A few placental mammals from the north made their homes in the tropical forests: jaguars became the main flesh eaters while spiny pocket mice and squirrels took over the rodent niches.

Of the northward invaders, opossums, porcupines and armadillos were the most successful though giant ground sloths and glyptodonts also survived in the north for a considerable time.

Other South American migrants travelled no further than tropical Central America, where most are still alive today.

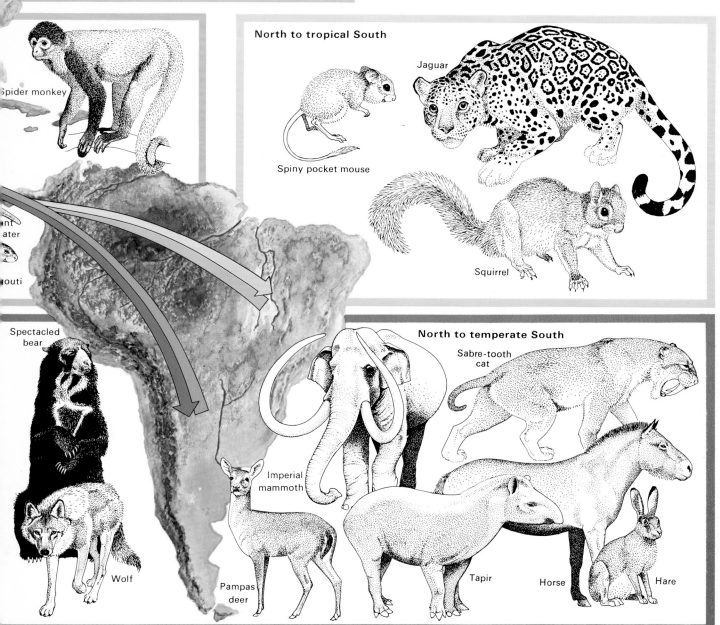

Spider monkey

North to tropical South

Spiny pocket mouse

Jaguar

Squirrel

Spectacled bear

Wolf

North to temperate South

Sabre-tooth cat

Imperial mammoth

Pampas deer

Tapir

Horse

Hare

The area where the tar pits formed was part of a flat plain. Rainwater does not drain away from pools of oily tar but collects on the surface in puddles. Animals looking for water would naturally be attracted and would gather to drink. The soft ground would sink under their feet and as they struggled to escape they would only be more deeply trapped.

Most of the plant eaters in the tar pits are bison, but there are also horses, pronghorns and camels. Various types of elephant are there (American mastodons and Imperial mammoths) and the giant ground sloths that had come from South America. The remarkable feature of the fossils is the proportion of flesh eaters to plant eaters: flesh eaters outnumber plant eaters by ten to one – the complete opposite to the normal situation where flesh eaters make up only about 3 % of the total weight of mammals in an area. If you imagine what happened when a large plant eater such as a mastodon was trapped it is easy to understand how this came about.

As the mastodon struggled, carnivores and scavengers would be attracted to the tar pit. The sabre-tooth cat *Smilodon* normally hunted young or weak elephants and would leap onto the exhausted animal and begin to tear off its flesh. This would attract scavengers such as the dire wolves and packs of them would soon appear to share the meat. Finally, vultures would wait their turn to pick the carcase clean.

Unfortunately, as soon as the flesh eaters stepped into the tarry pools they, too, were trapped fast. The only difference in their fate was that they were more likely to die of suffocation in the tar than from the bites of their fellow hunters and scavengers.

More than 2,000 skeletons of *Smilodon* have been recovered from the tar pits, with the remains of even more dire wolves. Judging by the fossils in the tar pits, *Smilodon* and the dire wolves seem the main flesh eaters of the time. Other cats such as Californian lions and pumas are preserved, but they are much less common. Among the dog family modern wolves and foxes have been found but, compared with the dire wolves they are rare.

When other fossil sites of the same period are studied, the picture looks quite different with dire wolves and sabre-tooths making up a much smaller fraction of the total number of carnivores. They were preserved in such large numbers in the tar pits simply because they were less intelligent than the other flesh eaters. Sabre-tooths hunted slow-moving prey and dire wolves were scavengers rather than active hunters. The biting cats and wolf packs hunted swifter animals and had to be more intelligent to survive. They are rare in the tarpits

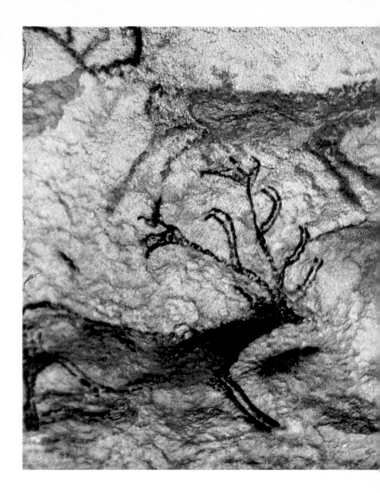

because they learned to recognize them as dangerous places and, in spite of the easy kills to be made there, they avoided them.

The Rancho la Brea tar pits are one of the few sites where the selection of animals preserved reflects their level of intelligence rather than simply where they lived.

After the Ice Ages

The last of the Ice Ages ended around 11,000 years ago. In North America many of the larger mammals such as mammoths, mastodons, giant sloths and glyptodonts died out at this time with their enemies the sabre-tooth cats and dire wolves. Many ruminants had already been driven out by the invading bison and deer from Eurasia and now horses and camels also disappeared.

In Europe cave bears, cave lions, mammoths and woolly rhinoceroses, all specially adapted to the cold climate, had all vanished by the end of the Pleistocene. Many other mammals had migrated south to the warmer tropical lands. As the ice finally retreated and the

Cro-Magnon man decorated the walls of caves with paintings of the animals among which he lived. We do not know why they were made but it is thought that they were connected with the people's system of religious beliefs. Cro-Magnon people lived by hunting so it is natural to assume that their religion would have been concerned with this way of life. By making a painting of an animal they may have believed that they were in some way capturing its spirit. Later, during the hunt, they would know that it was already spiritually wounded and would hunt with more confidence.

climate became better, large game animals such as the giant deer also died out.

Curiously there had been no mass extinctions like this in the earlier parts of the Pleistocene when the climate had improved between glaciations. Now in North America 35 different genera of mammals were reduced to a mere 13 over the space of a few thousand years. A similar change took place in Europe and Asia. The only new feature in the environment was man. Now living in every continent except Antarctica and constantly improving his hunting techniques, he had a disastrous effect on the delicate balance of the animal world.

In Africa, which even today has a large variety of game animals, the change was not so noticeable. Some animals did die out but they were replaced by the pigs and bovoids, the cattle, antelopes, sheep and goats which had migrated from Europe in the colder periods. In Africa the animals still live much as they did in the Miocene – occupying many very specialized ecological

niches. Even there, however, there had been a sudden drop from 26 genera of mammals to 19. This happened before the end of the Ice Ages, about 100,000 years ago and coincides with the time when man invented the hand-axe.

Man's improved hunting skills had an immediate effect on the other mammals but the balance of nature was soon restored. Only three genera of game animals have become extinct in Africa in the last 100,000 years and since the beginning of settled agriculture there have been no periods of general catastrophe. No genera of large game animals have, in fact, died out during the past thousand years.

Mammals in Palaeolithic art

The record of the history of the mammals is based almost entirely on the chance preservation of teeth and bones. This can never give a total picture of animal life in the past, for fossils are usually preserved only in places where sediments are being deposited – in rivers, swamps and estuaries. Animals preserved in other ways, like the mammoths and rhinoceroses frozen in the permafrost, are unusual and again are only a small selection of the animals living at the time.

About 20,000 years ago a new source of information on the animals of the past became available. In many places in Europe, but especially in Spain and France, Cro-Magnon man recorded the animals among which he lived on the walls of his caves. The paintings show details of animals that are now extinct and are the only really reliable evidence for making accurate, life-like reconstructions.

Next page: Los Angeles stands where pools of sticky tar once trapped thousands of animals in the Pleistocene. Fossils found there include imperial mammoths (1), sabre-tooth cats (*Smilodon* 2), dire wolves (3) horses (4), giant ground sloths (5), vultures (6) and pronghorns (7). More intelligent mammals including man (8) were rarely trapped.

The First Domestic Animals

Man had already become the dominant animal during the Ice Ages but when the last cold period finally ended he began a way of life that was to change the overall pattern of mammal life fundamentally.

As the ice caps melted and the climate improved, warm temperate grasslands spread in the Middle East and Asia Minor, in particular along what has become known as the Fertile Crescent, running from Iran through Iraq to Turkey, Syria, Palestine and Egypt.

Among the most significant plants were wild barley and wheat. Flocks of wild sheep and goats flourished in this environment. Wild sheep lived from Turkey eastwards to both the north and south of the Himalayas; goats were not so widespread, extending only as far as Afghanistan. In marked contrast pigs and cattle were found in a broad belt stretching from the Atlantic coast to the Pacific, including the coast of North Africa.

Sheep, goats, pigs and cattle made up the bulk of the mammals after the Ice Age. All in their different ways were highly adaptable and could flourish in a variety of environments. They undoubtedly represented the peak of herbivore evolution and in their turn were preyed on by the two most advanced and adaptable of the carnivores: the dogs and the cats. These six mammals between them were able to occupy all the major ecological niches available.

The land, however, was also occupied by one other species: *Homo sapiens*, man. Throughout the Pleistocene man had been developing into a highly efficient hunter and had spread into all the countries of the world. With his ever-improving hunting techniques he had undoubtedly hastened the extinction of many of the larger mammals. Around the beginning of the Holocene or Recent period, 11,000 years ago, large animals such as elephants and mammoths were either extinct or rare and man could no longer rely on them for food. In western Asia the flocks of wild sheep and goats provided a source of food that could hold its own against man because they reproduced so rapidly. Man's first step towards domesticating food animals was to turn to hunting sheep and goats instead of the larger prey animals. He remained, however, a hunter gatherer, collecting fruits, nuts and grain and following the herds as they moved from feeding ground to feeding ground.

The dogs

The dog family are group hunters, hunting in much the same way as humans. They are also general scavengers whenever the opportunity arises. Even more importantly, they respond to the leader of their pack. These three characteristics of the dog make it easy to see how

the close association between man and dog began.

In the cultural period known as the Mesolithic wolves and wild dogs of various kinds were living in many places. There is evidence in both Europe and Palestine that wolves were tamed and it is likely that it is from wolves that domestic dogs were bred.

Wolves and other wild dogs probably scavenged around human camps. If the adults were killed, captured cubs would be kept and gradually trained to become an important part of the hunting team – following the lead of a human being instead of a dominant pack leader. From the earliest times man must have known that wolves were skilful hunters and in time they were used to track down wounded animals and to drive prey into traps. Both dogs and humans benefited from their cooperation: the humans gained an efficient and swift moving hunter, the dogs were sure of a reliable and regular supply of food.

The beginning of agriculture

The Neolithic Revolution was the period when human beings established permanent settlements and cultivated plants such as barley and wheat, used for making gruel and beer. This was the beginning of agriculture. Up to this time man had followed wild sheep, goats, pigs and cattle; now he began to herd them himself.

The earliest records of domesticated sheep come from Iran, in deposits 10,500 years old. Where bones of sheep and goats that have been eaten are mainly from young or old animals, it shows they came from a wild herd where the mature and more active animals escape and only the weaker are killed. When the bones are almost entirely of young adults that have just reached their

Cats remain fierce hunters. Taking advantage of human shelter and food, they find their natural prey among the mice and small birds that live alongside man.

Roping cattle for branding. Beef cattle often live semi-wild, roaming over large areas of prairie. Fast, manoeuvrable horses are still more practical than machines when the cattle have to be herded and controlled.

Labrador tracking in the snow. Many domestic dogs are still working animals, used for hunting, protecting man's possessions and for controlling livestock on the farm.

maximum size, it proves they were from a domesticated herd. There is no advantage keeping them alive longer as they would not grow any larger.

Selective breeding to produce more suitable animals must have begun early. The large rams that dominated the wild herds were quickly killed off because they were too dangerously aggressive. The smaller rams which, in the wild, stood little chance of breeding, were used for siring the ewes. As a result, flocks of smaller, docile animals evolved.

Dogs were still used for hunting but were now also trained to herd sheep and to guard man's homes. In the wild, dogs defend their territory fiercely and they do the same when they are domesticated. Before long they had become more deeply involved than ever with man.

The first evidence that goats were domesticated also comes from Iran, about 1,500 years later than the sheep. For some reason we do not understand, the horns of both sheep and goats gradually change shape as they become domesticated. This, with the fact that the whole animal becomes smaller, gives a clue to when they were first tamed. Throughout the Middle East all the known villages had their own flocks of sheep and goats by about 9,000 years ago.

Wild pigs were domesticated in Turkey about 9,000 years ago and cattle, also in Turkey, about 500 years later. Cattle were the last animals to be domesticated, probably because the wild aurochs from which they decended were so large and fierce.

Agriculture and settled human communities seem to have first begun in the Middle East and the region is often called the cradle of modern civilization. However, man was domesticating other animals in many other parts of the world at much the same time. In China the banded pig was domesticated during the Neolithic. In North America the first record of domesticated dogs is from a site in Idaho over 10,000 years old. In South America dogs were also tamed and domesticated guinea pigs were being kept about 8,000 years ago. By 5,500 years ago man had domesticated the llama.

In Asia, cattle were domesticated between 4,000 and 5,000 years ago – the water buffalo in Pakistan and the banteng in Thailand. Five thousand years ago several ungulates were being used as pack animals and for transport in general – horses in the Ukraine, camels in Arabia and central Asia and in Egypt the ass. The onager was used to draw wheeled vehicles in Mesopotamia as long as 5,000 years ago. Curiously, the first evidence of the domestication of the reindeer, which had been part of man's staple diet during the Ice Ages, comes from Siberia only 3,000 years ago.

The cats

One of the last mammals to become domesticated was the cat, which is recorded by the ancient Egyptians about 3,600 years ago. When humans began to store grain and other food in their settlements, rodent pests such as rats and mice increased. This in turn would have attracted their natural predators, the small cats. Man tolerated, even encouraged these silent, stealthy hunters. The cat has in many ways remained much more independent than the dog but both owe much of their success to intelligence and adaptability. The cat's association with man is more one-sided than the dog's: they accept food and shelter but continue to hunt only their natural prey – and only when they want to.

The Future

The recent history of the mammals reads like a catalogue of disaster. Species after species has been driven to the brink of extinction through the deliberate activities of man. In many cases a mere handful of individual animals has managed to survive only because no-one knew where they were.

Species such as the American bison, which are calculated to have numbered some 60 million animals in the early 1700s were reduced to a few hundreds by the end of the last century. The sea otter of the Californian coast was hunted to extinction for its pelt and was believed extinct until a few were re-discovered in 1938. The Tasmanian marsupial wolf has not been seen for many years but is believed to be just surviving. The large cats such as the leopard and the cheetah have been almost totally destroyed for the sake of their fashionable furs. Other animals such as the pampas deer and the giant armadillo are on the verge of extinction because their natural habitat is being destroyed to make way for agricultural development. Polar bears, walruses and the monk seal were once very numerous but have been hunted until their future, too, is in the balance. They are now protected by international agreement and this may give them a chance to recover.

The history of the whaling industry provides one of the worst examples of human short-sightedness, greed and stupidity in the management of a natural resource. The number of large whalebone whales that can be killed without endangering the species has been known for many years, but the whaling nations, especially the Soviet Union and Japan, have always insisted on cropping more. Although it has been scientifically proved that the whales will become extinct if catches are not limited, advice, international pressures, even the rulings of the International Whaling Commission, are ignored. No other animal and no machine can convert the vast food resources of the ocean's plankton so efficiently into meat and oil that man can use. By destroying the whales we will lose, perhaps for ever, a valuable and economic source of food.

In many parts of the world there are extensive game parks and wildlife reserves which hold out hope for many species. Many are in developing countries where they are an important part of the tourist industry. This means that they are very dependent on the richer nations, for it is from these that most tourists come.

Even in the game parks the animals are not always safe. Poachers kill vast numbers of elephants for the ivory of their tusks. The largest tusks belong to the female leaders of the herds and these are naturally the most valuable. Left without a leader, the herds of elephants

Today people are the dominant mammals and their activities affect all other living things. The animals most likely to survive in the Age of Man are those which we farm, our companions the cats and dogs and those which can adapt successfully to an urban environment.

cause considerable damage to crops – and this, of course, leads to further killings.

Many people do now realize that species are in danger and the World Wildlife Fund, the International Union for the Conservation of Nature and many national Fauna Preservation societies are trying hard to save the world's wildlife heritage. They provide funds and experts to study animals in their environments and both governments and the public are gradually being brought to realize that animals need protection from indiscriminate slaughter.

Once zoos were happy to destroy several rare animals in order to obtain one living specimen but now, fortunately, this is becoming a thing of the past. Instead, special environments are preserved so that animals adapted to them can survive in their natural surroundings. In places where a species has died out, it may be reintroduced from elsewhere.

Man himself, the most adaptable of all mammals, will almost certainly survive. Whether we live in a world

Hedgehogs (above) have lived almost unchanged for over 60 million years and there is little doubt that they will continue to flourish. Foxes (right), house mice (below, right) and small hunters such as stoats (below) are all able to make use of human living habits, turning our food stores (and our wastefulness) to their own advantage.

with a wide variety of animals and plant life is up to us. The fate of most mammals is in our hands and their chance of survival depends very much on how useful they are to us. Which, then are most likely to continue?

Food animals are the most secure and new species are now being farmed as well as the familiar cattle, sheep, goats and pigs. In part of Nigeria the fruit bat *Eidolon* has been regularly cropped for several years. The giant rat *Cricetomys* and Maxwell's duiker have also been domesticated there. In North America herds of African eland are bred for meat and in the Soviet Union flocks of saiga antelopes are commercially farmed. Farming wild animals in this way provides a new supply of protein and, unlike game parks, does not depend on a steady stream of rich tourists for its success.

Factory farming, on the other hand, turns animals into little more than food processors and machines may eventually be made to convert plants to animal protein. Hopefully there will always be a market for 'free range' animal meat.

Two mammals that seem likely to continue as long as man himself are his two companions, the cat and the dog. Another group that depend entirely on man for their food are the rats and mice. These flourished throughout the Pleistocene and show no signs of declining. They are pests in both the countryside and in cities and although we continually try to get rid of them, we have finally come to accept them as part of the modern scene.

The last group of mammals that seem most likely to survive are the unspecialized opportunists. Inconspicuous animals such as hedgehogs and shrews have lived almost unchanged since the beginning of the Tertiary period 64 million years ago, perhaps for even longer. The success of foxes and opossums in cities shows that there are still mammals capable of adapting to the new environments made by man.

Man's growing feeling of responsibility towards his fellow creatures, together with the adaptability of many species proves that mammals do indeed have a future.

Glossary

AMPHIBIAN A vertebrate which lives in the water as a gill-breathing tadpole when young and develops into a land-dwelling, lung-breathing adult.

ARCHOSAUR A major group of reptiles, originally semi-aquatic, which included the dinosaurs, pterosaurs, birds and crocodiles. Crocodiles are the only remaining archosaurs as birds are now classed separately.

BROWSER A plant-eating animal that feeds on the leaves of trees and bushes, not on grasses.

CARNIVORE A flesh-eating animal, e.g. a lion.

CONGLOMERATE Sedimentary rock made up of pebbles.

CONVERGENT EVOLUTION This is said to have taken place when two unrelated animals adapt in the same way to similar environments.

CUSP A sharp, conical point on the biting surface of a tooth.

ECOLOGICAL NICHE An animal's particular place in nature, including its habitat, feeding habits, size, time of day it is active etc.

ECTOTHERM (cold blooded) An animal that controls its internal temperature by taking advantage of sun and shade to heat up or cool down.

EDENTATE (literally without teeth) A group of South American animals, armadilloes, sloths and anteaters. In spite of their name, all except the anteaters do have teeth.

EMBRYO Young developing inside an egg or mother animal.

ENDOTHERM (warm blooded) An animal that controls its internal temperature by mechanisms within its own body.

FOOD CHAIN The way in which animals and plants are linked to one another as food, e.g. flesh-eating animals eat plant-eating animals which in turn eat plants.

FOSSIL Preserved evidence of living things from the past – petrified or mineralized bones, parts of plants, even footprints may be preserved.

GENUS (pl. genera) A grouping of similar species within a zoological family.

GRAZER A plant-eating animal that feeds on grasses.

HABITAT Where an animal lives, its environment.

HERBIVORE A plant-eating animal; includes both grazers and browsers.

INSECTIVORE An insect-eating animal.

MAMMAL Warm-blooded, usually furry vertebrate which suckles its young.

MARSUPIAL Pouched mammal that bears young at early stage of development and retains it in a pouch until it is able to be independent. A few marsupials such as opossums do not have pouches but keep their young on their back.

MEMBRANE A thin sheet of tissue.

METABOLIC RATE The rate at which food is burned up in the body to produce energy.

MONOTREME Primitive egg-laying mammal.

NATURAL SELECTION The process by which the animals most suited to their particular environment survive longest and therefore breed more, passing on their special characteristics. Survival of the fittest.

PARALLEL EVOLUTION The way in which closely related animals gradually evolve in similar (or parallel) ways in response to environmental change.

PREDATOR An animal that preys on others for food.

PRIMATE Order of mammals which includes bushbabies, monkeys, apes and man.

REPTILE An egg-laying, scaly, air-breathing, cold-blooded vertebrate.

RUMINANT A group of animals with stomachs specially adapted for processing plant food. A cud-chewer.

SCAVENGER An animal that lives by eating up the remains of other animals' kills.

SEDIMENT (of rock) Broken-down rocks carried in small grains by rivers, wind etc.

SEDIMENTARY ROCK Rock formed from layers of sediments, e.g. sandstone, conglomerates, clays.

SEMI-AQUATIC An animal that spends its life partly in water, partly on land, not necessarily an amphibian.

SPECIES A distinct group of animals which can mate with one another and produce fertile young; a sub-division of genus.

TUBERCULE A small lump or swelling.

VERTEBRA Individual bone of backbone.

VERTEBRATE Animal with a backbone.

Bibliography

BROTHWELL, D ed.: *The Rise of Man*, Sampson Low, Maidenhead, 1976

CARRINGTON, R: *Elephants*, Chatto and Windus, London, 1958; Penguin Books, Harmondsworth, 1961

CARRINGTON, R: *The Mammals*, Time Life, New York, 1965

CLARK HOWELL, F: *Early Man*, Time Life, New York, 1966

COLBERT, E H: *Evolution of the Vertebrates*, 2nd ed. Science Editions, New York, 1961

HALLAM, A ed.: *Planet Earth*, Elsevier-Phaidon, Oxford, 1977

HALSTEAD, L B: *The Pattern of Vertebrate Evolution*, Oliver and Boyd, Edinburgh, 1969

HALSTEAD, L B: *Vertebrate Hard Tissues*, Wykeham Publications, London, 1974

HALSTEAD, L B: *The Evolution and Ecology of the Dinosaurs*, Peter Lowe, London, 1975

HALSTEAD, L B and MIDDLETON, J A: *Bare Bones, an exploration in art and science*, Oliver and Boyd, Edinburgh, 1972

HAMILTON, W R: *The History of Mammals*, British Museum (Natural History), London, 1972

KURTÉN, B: *The Age of Mammals*, Weidenfeld and Nicolson, London, 1971

LAVOCAT, R: *Histoire des Mammifères*, Editions du Seuil, Paris, 1967

ROMER, A S: *Vertebrate Paleontology*, 3rd ed., Chicago University Press, Chicago, 1966

SCOTT, W B: *A History of Land Mammals in the Western Hemisphere*, 2nd ed., Macmillan, New York, 1937

SIMONS, E L: *Primate Evolution*, Macmillan, New York, 1972

SIMPSON, G G: *Horses*, Oxford University Press, New York, 1951

SPINAR, Z V and BURIAN, Z: *Life before Man*, Thames and Hudson, London, 1972

TIME-LIFE Editors: *The Missing Link*, 1973; *The First Men*, 1973; *The Neanderthals*, 1973; *Cro-Magnon Man*, 1973; *The First Farmers*, 1974. Time-Life, Amsterdam

WOOD, B: The Evolution of Early Man, Peter Lowe, London, 1976

YOUNG, J Z: *The Life of Vertebrates*, Clarendon Press, Oxford, 1962

Index

Acknowledgements

The author would like to thank Professor J. L. Albright and Miss Linda Barber for their kind help.

Artists
John Barber 34,36,41,42,46,48,50,62,80,82,84,96
Rudolf Britto 12,16,25,30,31,35,42,48,58,59,60,63,65,66,71,81,82,
85,86,89,90,99,102,104
Enzo Carretti 22,26
Giovanni Caselli 72-3, 86,88,89
David Etchell 17,18,22,27
Tom MacArthur endpapers
Malcolm McGregor 55,56,60,100
Sean Milne 30,66,83,87,92
Richard Neave 10-11,14-15,67
Lorenzo Orlandi 70,74
Sergio title page, 32,38,44,52,64,68,78,90,94,98,108
Tony Swift 24,102
Michael Woods endpapers, 14,21,28,35,40,55,73,75,77,97,104

Photographs
Heather Angel 13,111,113
Donald Baird 19
Biofotos 13
Robin Fletcher 10
Michael Fogden 17
Natural History Photographic Agency 12,110,11
Imitor 20,77
Natural Science Photos 12,20
Novosti Press Agency 20
Zefa 111,112,113

Index by Susan Abbott

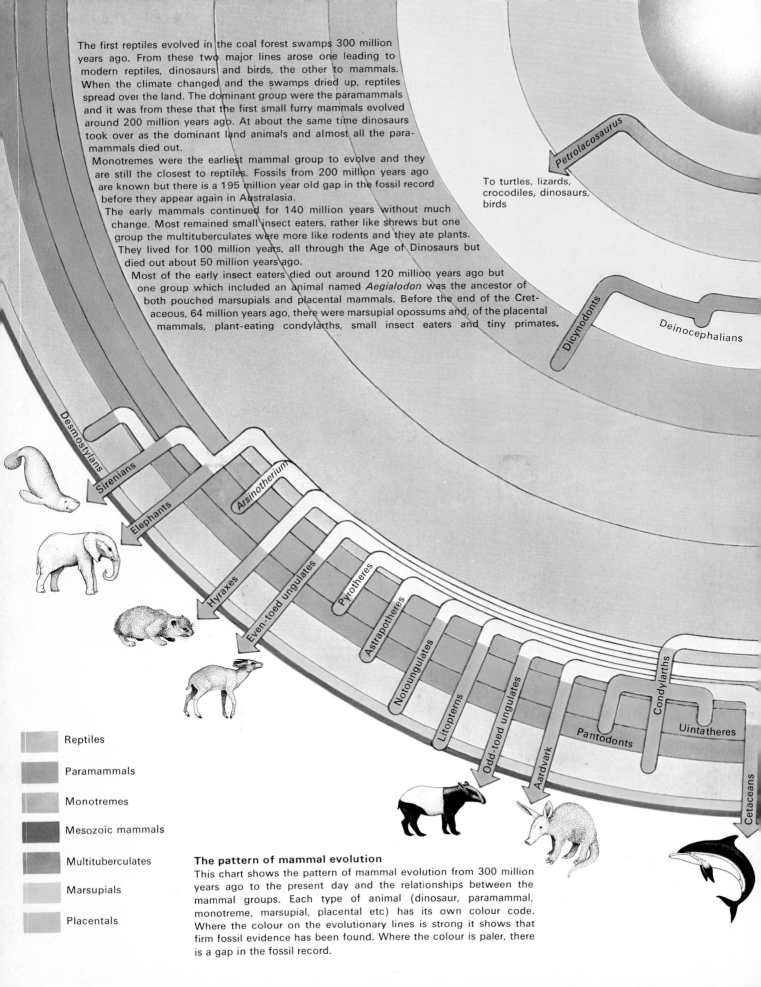

The first reptiles evolved in the coal forest swamps 300 million years ago. From these two major lines arose one leading to modern reptiles, dinosaurs and birds, the other to mammals. When the climate changed and the swamps dried up, reptiles spread over the land. The dominant group were the paramammals and it was from these that the first small furry mammals evolved around 200 million years ago. At about the same time dinosaurs took over as the dominant land animals and almost all the para-mammals died out.

Monotremes were the earliest mammal group to evolve and they are still the closest to reptiles. Fossils from 200 million years ago are known but there is a 195 million year old gap in the fossil record before they appear again in Australasia.

The early mammals continued for 140 million years without much change. Most remained small insect eaters, rather like shrews but one group the multituberculates were more like rodents and they ate plants. They lived for 100 million years, all through the Age of Dinosaurs but died out about 50 million years ago.

Most of the early insect eaters died out around 120 million years ago but one group which included an animal named *Aegialodon* was the ancestor of both pouched marsupials and placental mammals. Before the end of the Cretaceous, 64 million years ago, there were marsupial opossums and, of the placental mammals, plant-eating condylarths, small insect eaters and tiny primates.

Petrolacosaurus

To turtles, lizards, crocodiles, dinosaurs, birds

Dicynodonts

Deinocephalians

Desmostylians

Sirenians

Elephants

Arsinotherium

Hyraxes

Even-toed ungulates

Pyrotheres

Astrapotheres

Notoungulates

Litopterns

Odd-toed ungulates

Aardvark

Pantodonts

Condylarths

Uintatheres

Cetaceans

Reptiles

Paramammals

Monotremes

Mesozoic mammals

Multituberculates

Marsupials

Placentals

The pattern of mammal evolution
This chart shows the pattern of mammal evolution from 300 million years ago to the present day and the relationships between the mammal groups. Each type of animal (dinosaur, paramammal, monotreme, marsupial, placental etc) has its own colour code. Where the colour on the evolutionary lines is strong it shows that firm fossil evidence has been found. Where the colour is paler, there is a gap in the fossil record.